Investigating Geography C

JACKIE ARUNDALE · SUE BERMINGHAM · SIMON CHANDLER · CHRIS DURBIN
GREG HART · BOB JONES · LINDA KING · FRED MARTIN · DIANE SWIFT
SERIES EDITORS: KEITH GRIMWADE & CHRIS DURBIN

Hodder & Stoughton

A MEMBER OF THE HODDER HEADLINE GROUP

Acknowledgements

The publishers would like to thank the following individuals, institutions and companies for permission to reproduce copyright illustrations in this book:

© actionplus, 3.1tl; © AP Photo/Charles Rex Arbogast, 2.11; © AP Photo/Namas Bhojani, 2.12; © AP Photo/Sherwin Crasto, 3.23; © AP Photo/Eugene Hoshiko, 2.13, 3.22; © AP Photo/Dolores Ochoa, 3.21; AP Photo/Gorans Veljkovic, 3.1br; © Yann Arthus-Bertrand/Corbis, 4.4; Associated Press/AP, 1.21, 2.7, 3.16, 3.17, 3.19, 3.30, 3.31, 5.18a, 6.13; Associated Press, EFE, 1.19; © Steve Austin, Papilio/Corbis, 5.21 (vole); © Paul Barton/Corbis, p4 1.1br; © Jonathan Blair/Corbis, 5.18c; Bluetown History & Arts Centre, 4.2; Joerg Boethling/Still Pictures, 5.6iv; Bright Ideas, 5.29 (logo); Simon Chandler, 5.24, 5.25, 5.26; © Lloyd Cluff/Corbis, 3.29; Corel, p7 (flag), 1.4, 1.13 (flags), p38 (logo), p40br, p52 (flags), p61 (flag), p82 (Cardiff); © Philip James Corwin/Corbis, 6.24; © James Davis, Eye Ubiquitous/Corbis, 5.20c; Cartoon from Thin Black Lines – Political Cartoons and Development Education (DEC), 1.29; Digital Vision, 3.36b; Mark Edwards/Still Pictures, 5.4, 5.6ii, 5.18b; Eyewire, p40t; © Friends of the Earth, 6.43; © Colin Garrett, Milepost 92¹/₂/Corbis, 6.29; GRAFF/Cartoonists & Writers Syndicate/cartoonweb.com, 2.6; Hodder Arnold, 6.31 (landfill); © Robert Holmes/Corbis, 4.7; © Hulton-Deutsch Collection/Corbis, 1.17; Image Boss, pp16-17 (shadows and bg); Ingram Publishing, p36 (man and computer), p81bg, 5.1, 5.21 (mushroom, fox); Bob Jones, 4.1, 4.3, 4.5, 4.8, 4.13, 4.14, 4.15, 4.18, 4.19, 4.22, 4.23, 4.24, 4.26, 4.27, 4.28, 4.29, 4.30, 4.31, 4.32, 4.34, 4.35, 4.36, 4.37, 4.38, 4.39, 4.40, 4.41, 4.43, 4.44, 4.45, 4.46, 4.47, 4.48, 4.49, 4.51, 6.2, 6.14, 6.16, 6.17, 6.18, 6.19, 6.21, 6.23, 6.25, 6.27, 6.28, 6.31 (all except landfill), 6.32, 6.33, 6.35, 6.36, 6.37, 6.38, 6.39; University Library, Keele/The Warrillow Collection, 6.1; © Frank Lane Picture Agency/Corbis, 5.21 (weasel); Life File/Graham Burns, 3.1tr; Life File/Xavier Catalan, 1.16; Life File/Emma Lee, p4 1.1tr, p5 1.1tl, p5 1.1cr, p5 1.1b, 1.35t; Life File/Barry Mayes, 6.10; Life File/Richard Powers, 6.11; Life File/Eddy Tan, 1.20; J. Freund/Still Pictures, 5.3; © Wolfgang Kaehler/Corbis, 5.20b; © George D. Lepp/Corbis, 5.6iii; © Buddy Mays/Corbis, 5.19, 5.20a; © Amos Nachoum/Corbis, 5.28; Nomad, p7 (tv), pp16-17 (eyes), 3.2b, 4.50bg, 6.9a; Christine Osborne/Ecoscene, 2.18; Photodisc p4 1.1l, p7 (seaside and night), 1.13bg, 1.15bg, 1.35b, 2.1, p29bg, p35 (monitor), p36 (women), p40cr, p40bl, p40bc, 3.2t, 3.2c, 3.5, 3.15, 3.35bg, 3.36t, 3.36c, 4.6, 4.20, p82 (Mexico), 5.5, 5.6i, 5.8, 5.10, 5.21 (rabbit, insect, small bird, vegetation), 5.23, 5.31, 6.5, p106, p107, 6.41; Gerry Quinn g.quinn@planetquinn.com, 2.9; Adrian Raeside/www.raesidecartoon.com, 2.6; © Roger Ressmeyer/Corbis, 3.11a; © Galen Rowell/Corbis, p5 1.1cl; © Kevin Schafer/Corbis, 3.10a; © Copyright by the local government of Schlema, 6.22, 6.26; © Paul A. Souders/Corbis, 3.4, 5.9; © Studio Burie, Netherlands, 2.2a, 2.2b; © W.T. Sullivan III/Science Photo Library, 2.3; © UNDP, 2.15; US Geological Survey, 3.1c; © Michael Yamashita/Corbis, 3.28.

 (l = left; r = right; t = top; b = bottom; c = centre; bg = background)

The publishers would also like to thank the following for permission to reproduce material in this book: extract adapted from Insight Guide Spain, ©2002 Apa Publications GmbH & Co Verlag KG (Singapore Branch); UN for various extracts in chapter 2; the extract from the Collins Longman Atlas (redrawn) © Bartholomew Ltd 2002. Reproduced by permission of Harper Collins Publishers www.bartholomewmaps.com; This product includes mapping data licensed from Ordnance Survey® with permission of the Controller of Her Majesty's Stationery Office. © Crown copyright 2002. All rights reserved. Licence number 100019872.

Note about the Internet links in the book. The user should be aware that URLs or web addresses change regularly. Every effort has been made to ensure the accuracy of the URLs provided in this book on going to press. It is inevitable, however, that some will change. It is sometimes possible to find a relocated webpage by just typing in the address of the home page for a website in the URL window of your browser.

Orders: please contact Bookpoint Ltd, 130 Milton Park, Abingdon, Oxon OX14 4SB. Telephone: (44) 01235 827720. Fax: (44) 01235 400454. Lines are open from 9.00 - 6.00, Monday to Saturday, with a 24 hour message answering service. You can also order through our website www.hodderheadline.co.uk.

British Library Cataloguing in Publication Data
A catalogue record for this title is available from the British Library

ISBN 0 340 80377 0

First Published 2002
Impression number   10 9 8 7 6 5 4 3 2
Year   2008 2007 2006 2005 2004

Cover photo from Photodisc.
Designed and typeset by Nomad Graphique.
Printed in Italy for Hodder & Stoughton Educational, a division of Hodder Headline, 338 Euston Road, London NW1 3BH.

Investigating Geography C

Contents

1 Spain

Spain or not?

▽ Figure 1.1 ▷

Activities

1 For each image in Figure 1.1, decide whether or not you think it shows Spain. Write a sentence to explain your decision.

2 For each image find out how many people in your class thought that it was of Spain.

3 Did any images catch out most of the class? Why?

4 Think about a country that you know well. Which seven images would you choose to represent that country? Why would you choose those images?

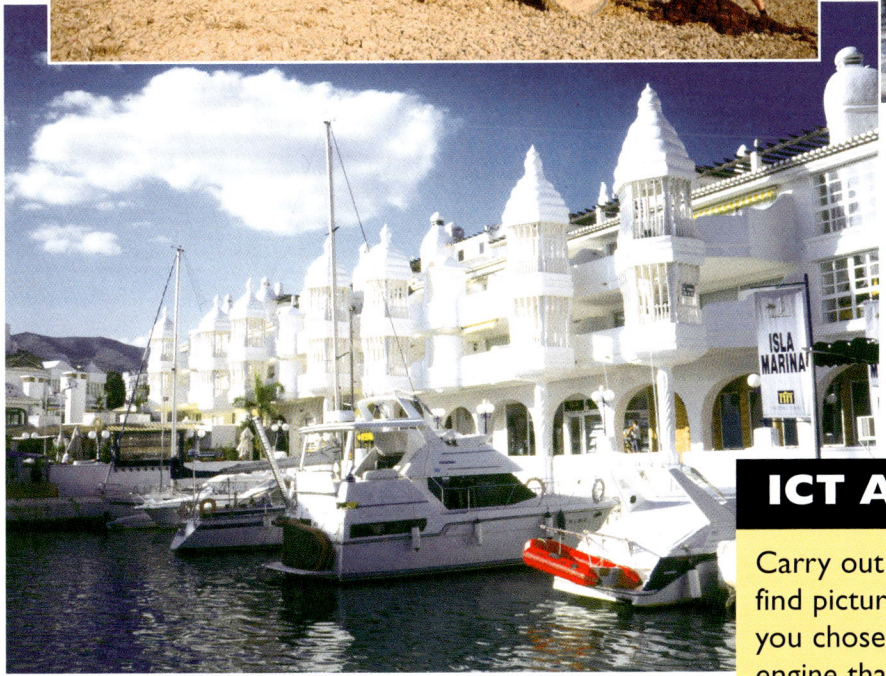

ICT Activity

Carry out a search on the Internet to find pictures for some of the images you chose for question 4. Use a search engine that has an image search facility, such as www.google.com.

What are your first thoughts?

In this chapter we are going to investigate Spain. In Book B, chapter 2 on development explored what the terms **nation**, **country** and **state** can mean. It stated that 'a nation can be said to be a group of people who share a common culture, history or language and have a feeling of national unity'. You are going to investigate whether or not Spain is a nation. To do this you will be using geographical information at the **national**, **regional**, **local** and **personal** scale. You will need to be a critical reader and user of the information that you receive.

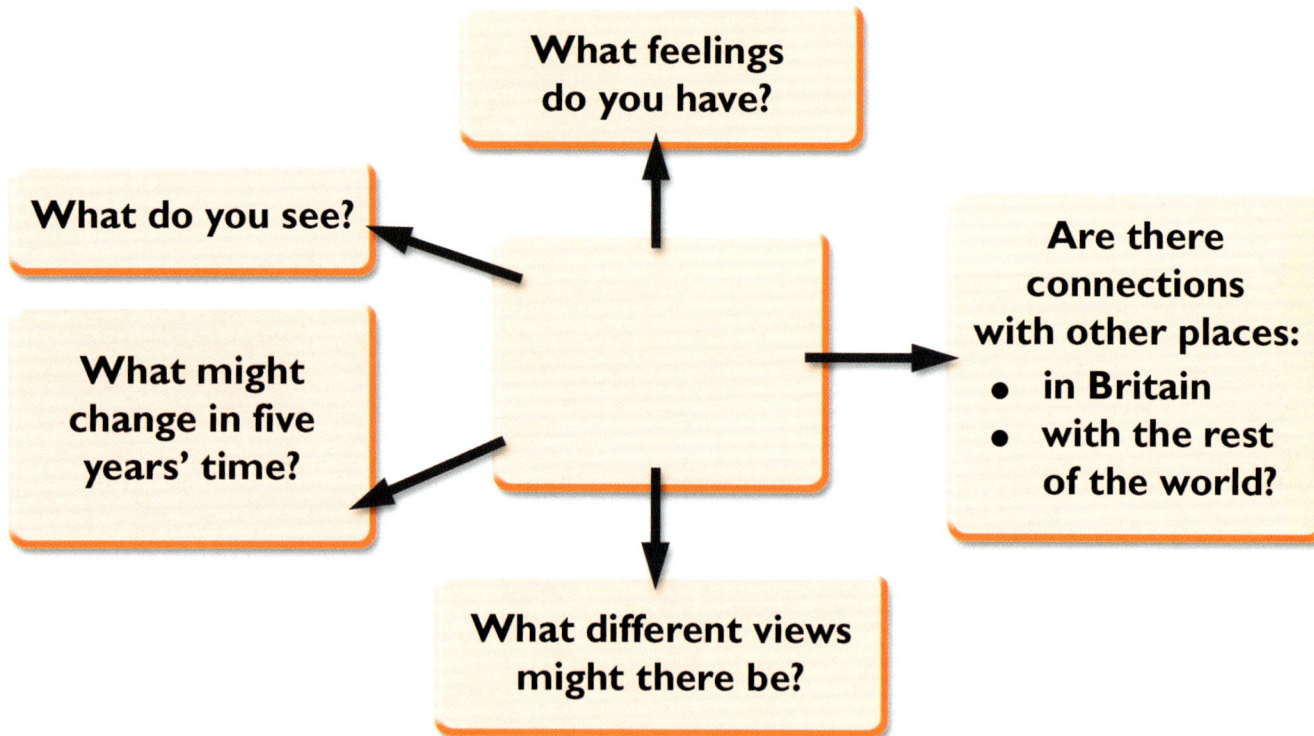

What feelings do you have?

What do you see?

What might change in five years' time?

Are there connections with other places:
- **in Britain**
- **with the rest of the world?**

What different views might there be?

△ **Figure 1.2** Learning from an Image frame

Activities

1 Select one of the photographs on pages 4 and 5.

2 On a large sheet of A3 paper, copy the Learning from an Image frame (Figure 1.2).

3 Complete the frame for your chosen photograph.

4 Display your frame with the others around the classroom. You might want to group together the frames that are of the same photograph.

5 Make a note of the things that you have learnt by reading other people's frames.

6 Think for a moment about what you already know about Spain. Make a note of the words, images and feelings that come to mind.

7 Now think about how you have found out that information. Again make a list.

8 Keep both lists safely; you will need to refer to them again.

How do you develop a viewpoint?

A Sense of Place

Geographical information is plentiful. In fact there is so much of it that quite often we need to select from it and simplify it. The information that we have about Spain will be partial. It will not be complete. It will also have been written or produced from a particular viewpoint or for a particular purpose. This view or purpose may well have influenced the content and type of information that we see. We have to be careful how we use it. We have to be careful what we understand from it. So we need to think about geographical knowledge. These three stages might be helpful.

Stage A – We are informed

- We are informed by partial images and information that come to us through the TV screen and computer monitor, newspapers, textbooks, teachers, family and friends.
- All these sources and people will have put their own interpretation on the information that they choose to share with us.

Stage B – Experiences inform our knowledge

- Our experience (or lack of experience) may have shaped how we feel about information.
- Our experience and feelings may change.
- Our information is usually incomplete.
- Further information or evidence can change what we know.

Stage C – Knowledge is always developing

- What we know about Spain is shaped by our view (perspective).
- Our perspective can change as a result of learning more through our experiences, what we see, hear and can touch, and, through our inner thought processes, how we think and feel about our various experiences.

△ **Figure 1.3** Sense of Place frame

Activities

1 Look back at the work you did for questions 1–8 on page 6. Use the Sense of Place frame above. Think about how you used and interpreted the geographical information that you have used from the photographs. Is there anything that you want to add?

What do you know about Spain?

1 Spend three minutes studying the relief map of Spain opposite. Do not write anything down; do not talk or discuss what you see with anyone. At the end of the three minutes, close this book and put it out of sight.

2 Your teacher will ask you to work with a partner. Spend one minute each, telling the other about what you remember from the map.

3 Look again at the map opposite. Using this as a guide, draw an outline map of Spain. Label the map with all the things that you and your partner have remembered.

4 Use a different colour to add anything that you and your partner didn't remember, but think is important. While you are talking about some of the things you have remembered, you might like to use some Spanish words from Figure 1.4.

5 Use an atlas. Find all the maps in the atlas which feature Spain. What extra information do these maps give you? Make a list under the headings: **Environment**, **Economic**, **Cultural**.

6 Can you use this information to add a paragraph to the description started on page 9? You may wish to use the headings from question 5 to help you.

Useful words

These phrases may be useful:

English	Spanish
hello	hola
goodbye	adiós
good morning	buenos días
good afternoon	buenas tardes

English	Spanish
yes/no	si/no
please	por favor
thank you (very much)	(muchas) gracias
where is...?	¿donde está...?

These phrases m...

English	Spanish
excuse me, I don't understand	perdón, no entiendo
could you please speak more slowly?	¿podría hablar más despacio, por favor?

English	Spanish
I am/my name is...	soy/me llamo...
I am from...	soy de...
I am English	soy inglés
what is your name?	¿como se llama usted?

English	Spanish
how are you?	¿como está?
very well	muy bien

△ **Figure 1.4** Useful Spanish words and phrases

Spain occupies 85% of the Iberian **peninsular**. 88% of its boundary is with water. It is bordered in the north by the Bay of Biscay, France and Andorra, on the east by the Mediterranean, on the south by the Mediterranean and the Atlantic Ocean, and on the west by Portugal and the Atlantic Ocean. The small self-governing British colony of Gibraltar is on the southern extremity of Spain. The Balearic Islands in the Mediterranean and the Canary Islands in the Atlantic Ocean off the coast of Africa are governed as provinces of Spain.

Spain has a population of 40 037 995 (2001). The overall population density is about 79 people per square kilometre. Spain is increasingly urban with 77% of the population living in towns and cities.

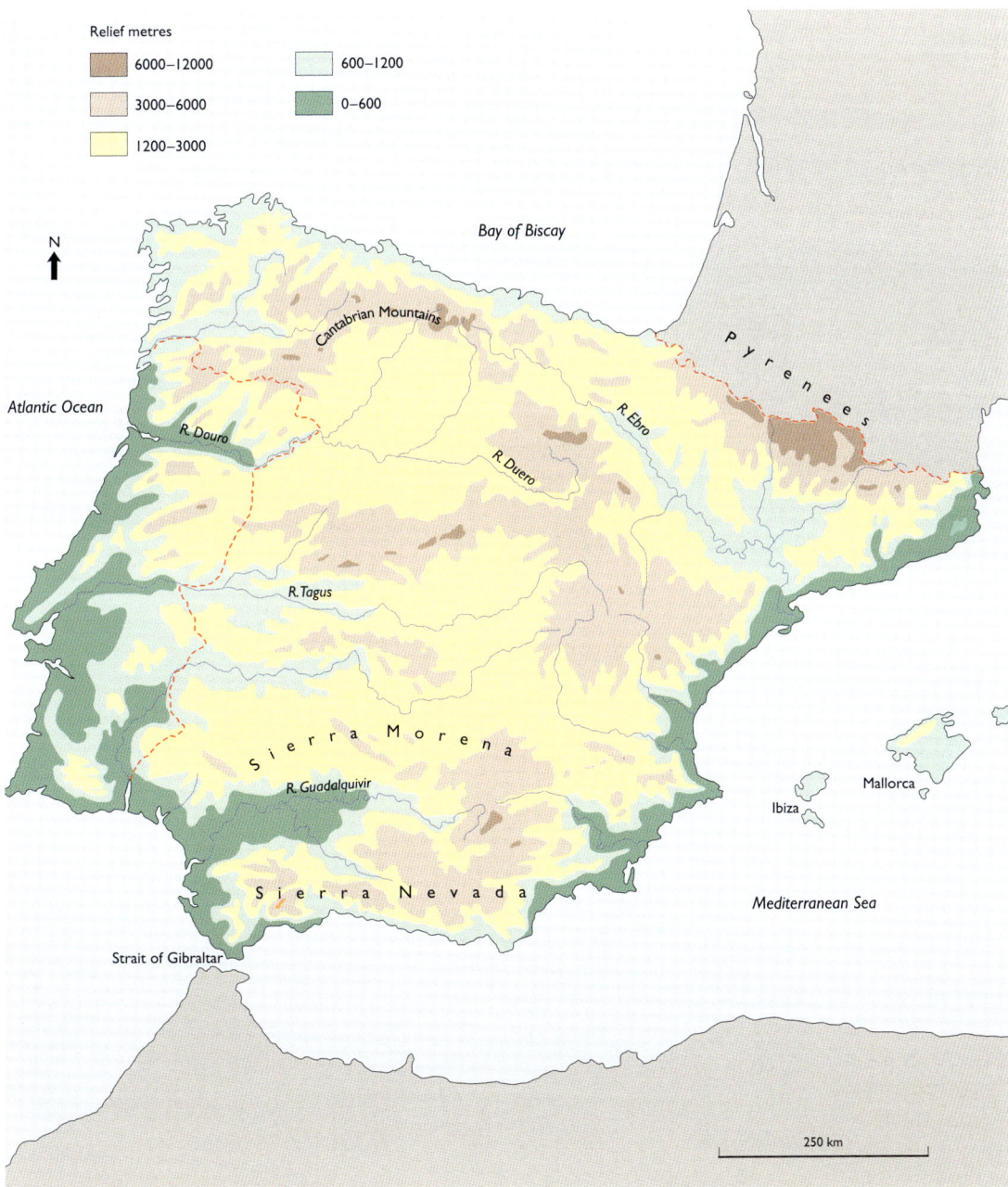

Relief metres

6000–12000	600–1200
3000–6000	0–600
1200–3000	

N

Bay of Biscay

Cantabrian Mountains

P y r e n e e s

Atlantic Ocean

R. Ebro

R. Douro

R. Duero

R. Tagus

Sierra Morena

R. Guadalquivir

Mallorca

Ibiza

Sierra Nevada

Mediterranean Sea

Strait of Gibraltar

250 km

◁ **Figure 1.5** Relief map of Spain redrawn from © Bartholomew Ltd 2002. Reproduced by permission of Harper Collins Publishers www.bartholomewmaps.com

How to make sense of Spain?

So far in this chapter, we have used words, pictures and maps to find out about Spain. We are now going to investigate some statistics about Spain's population. One way to test our developing knowledge is by comparison, so data for four other countries has been included.

Norway has been selected because in 2001 it ranked number 1 in the UN's Human Development Report as being the best country in the world to live in. The United States, Japan and United Kingdom will also make interesting comparisons.

Country / year	1975	1999	2015
Norway	4.0	4.4	4.7
United States	220.2	280.4	321.2
Japan	111.5	126.8	127.5
United Kingdom	56.2	59.3	60.6
Spain	35.6	39.9	39.0

△ **Figure 1.6** Total population – millions

Activities

Foundation

1 Draw a divided bar graph to show each of the four countries, total population for 1975, 1999 and the predicted population for 2015. The divided bar graph for Norway would look something like this.

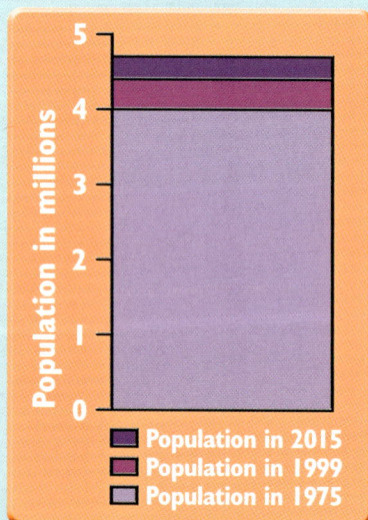

△ **Figure 1.7** Norway: bar graph showing population in millions, 1975–2015

Target

2 Look at the figures presented in the table in Figure 1.6 and using a blank world map outline, **annotate** the map to illustrate the population figures for Spain and the four countries chosen for comparison. Write a paragraph to describe your map.

3 **Population density** is the figure calculated to tell you on average how many people live in each square kilometre of a country. From your world map which do you predict will be the least densely populated country and which will be the most densely populated country in 2015?

Extension

4 Use an atlas to find relief maps of the five countries listed in Figure 1.6. Remember that there is a relief map of Spain on page 9.

5 On a blank world map outline, annotate the map with as much detail as possible about the total population for each country and how you think that population might be distributed in the country.

6 In 2001 Spain had a population density of 79 people per square kilometre. How do you think Spain compares with the other four countries?

7 What might be some of the strengths and weaknesses of a population density statistic?

Country / year	1975	1999	2015
Norway	68.2	75.1	80.1
United States	73.7	77.0	81.0
Japan	75.7	78.6	81.5
United Kingdom	88.7	89.4	90.8
Spain	69.6	77.4	81.3

△ **Figure 1.8** Urban population – as a percentage of total population

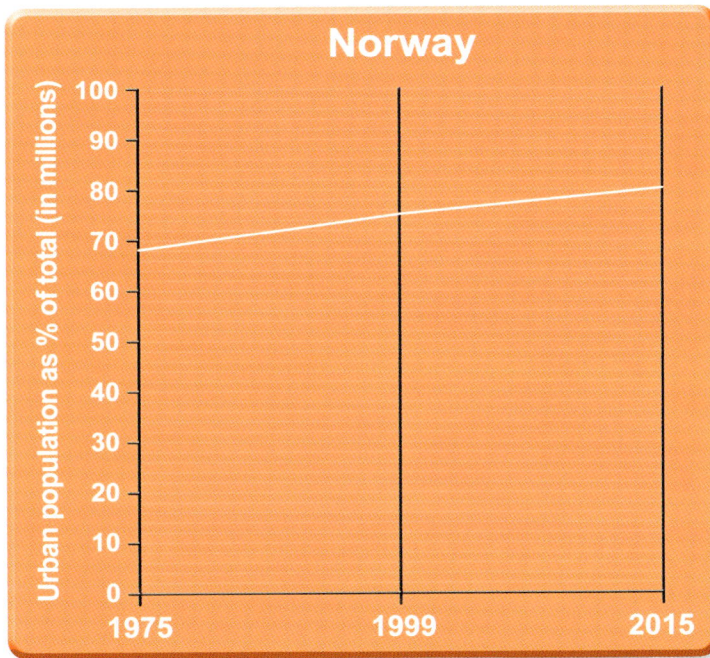

△ **Figure 1.9** Line graph to show Norway's urban population 1975–2015

Country / year	1999	2015
Norway	19.8	15.8
United States	21.9	18.7
Japan	14.9	13.3
United Kingdom	19.1	15.1
Spain	15.0	12.5

△ **Figure 1.10** Population aged 15 and under – as a percentage of total

Country / year	1999	2015
Norway	15.5	18.2
United States	12.3	14.4
Japan	16.7	25.8
United Kingdom	15.7	18.9
Spain	16.7	19.8

△ **Figure 1.11** Population aged 65 and above – as a percentage of total

Country / year	1999	2015
Norway		
United States		
Japan		
United Kingdom		
Spain		

◁ **Figure 1.12** Population aged between 15 and 65 – as a percentage of total

Activities

8 Figure 1.9 is a line graph that shows urban population as a percentage of population for Norway. Copy and complete the line graph to show the urban population for the other four countries in Figure 1.8.

9 Use an atlas. Which six cities might be Spain's largest in 2015?

10 By what percentage are they likely to have increased between 1999 and 2015?

11 What are some advantages and disadvantages of **urbanisation**?

12 Look at the Figures 1.8, 1.9 and 1.10. Copy and complete the table in Figure 1.12.

13 What do we mean by the dependent population? Why is it important to know this?

14 Using the information from all five tables on these two pages, write a paragraph to describe Spain's key population characteristics.

One Spain or several?

Las Españas

△ **Figure 1.13** Spanish national flag and flags of the regions

Spaniards often speak of their country as Las Españas – in the plural – reflecting the feeling that there is more than one Spain. The maps and statistics used to investigate Spain so far have given the impression of one nation. The reality is far more complex than this. One of the largest countries in Europe, Spain has a wide variety of climates and landscapes and also a mixture of people with different backgrounds. No one area nor any one Spaniard can be said to be typically Spanish.

The populations in several regions of Spain have kept a separate identity, culturally and linguistically. These include the Basques, who number about 2.1 million and live chiefly around the Bay of Biscay; the Galicians, about 2.5 million who live in the north-west of Spain; and the Catalans of eastern and north-eastern Spain.

Activities

1 Draw an outline map of Spain. Again you may wish to use the relief map on page 9 to help you. If you drew an outline map before, have a look to see if there are any improvements that you would like to make. Annotate your map to show 'three Spains in one'.

Places or spaces?

One of the things that geographers are particularly interested in is the way that places are created. This is normally a mixture of human and physical processes. In the rest of this chapter we are going to concentrate on a political aspect of the human processes that help make places unique. We are going to look deeper into how politics and power in Spain has affected where people live and how they feel about that place.

What are you?

If asked 'What are you?' most people will answer, 'I am English or American or French ...' and then proceed to specify, 'I come from Devon or Florida or Normandy...' ending up with 'My home town is Exeter or Miami or Cherbourg...'. The Spaniard, on the other hand, will reverse this order and start with 'I am from Denia' (most important), 'I am Valencian' (of secondary interest) and, if pressed further will admit 'Oh yes, I am Spanish'. The Spaniard's **individualism** is reflected in this peculiar identification with the local area rather than the country as a whole. Anyone coming from outside Spain is an *extranjero* but even a Spaniard from the next village will be referred to as a *forastero* (stranger). To this day, Spaniards retain this extreme individualism.

△ **Figure 1.14** An extract from *Culture Shock: A Guide to Customs and Etiquette, Spain* by Marie Louis Graff

Sense of identity

Sense of belonging

Understanding of others in a place

Personal benefits from places

△ **Figure 1.15** Sense of Belonging diagram

Activities

2 Read the passage on the left (Figure 1.14). Think for a moment about yourself and a place to which you feel that you belong. It may be the place that you live now, but equally it may not. It might be a place very different to the one where you are now.

3 Make a copy of the diagram above (Figure 1.15). Add notes about the place that you feel that you belong to.

4 Look carefully at your diagram. Try to use the labels to explain to someone why you are attached to that place. Your explanation is what geographers describe as a sense of place.

Spain – geography or history?

To explain Spain's geography we need to be aware of its history. To work towards an understanding of how people are distributed across the regions in Spain, it is important to understand why certain people feel such a sense of belonging to certain places.

A brief history of the people in Spain

The Iberian peninsula has been occupied for many millennia. The Basque people were the earliest identifiable group. The Romans conquered it in the first century BC. They kept control until firstly European barbarian tribes and then Muslims (Moors) from North Africa overran it in 711AD. Gradually the native Catholics fought back until in 1492 they unified the various kingdoms under one ruling family and ended religious tolerance. During the 16th century, Spain became the most powerful nation in Europe, mainly due to its overseas possessions, its powerful army and its silver bullion from America as well as its control of other areas of land. However, the 17th century saw military defeats, economic failure and poor rulers, which resulted in the loss of its European and overseas empire. Its steady political decline continued into the 20th century.

▷ **Figure 1.16** A Spanish galleon active in the 16th century, now a tourist attraction in Barcelona Maritime Museum

Spain's modern history

Spain was not directly involved in either the First or the Second World Wars but, in 1936, **civil war** broke out. This was caused by fears that there would be a socialist revolution in Spain. The Nationalists, backed by the military under General Franco, won. As a result, Spain became an isolated country, not joining the United Nations until 1955. Franco was a dictator; he did not devolve power. This was particularly hard to bear for some of the **regions**, for example the Basque country. During the 1960s and 70s, Spain changed into a modern industrial country with a growing tourist industry. Franco died in 1975 and was succeeded by King Juan Carlos. In 1977 Spain held its first elections since 1936 and is now a **democracy**. Each of its 17 regions rules itself; very different to Franco's rule.

What has Spain's history got to do with its geography?

The Basque country, Catalonia and Galicia consider themselves to be historic nationalities, and until Franco's rule, had remained independent from central government since the 16th century. In these regions local languages are spoken.

△ **Figure 1.17** Men in Madrid clear rubble following bombing by nationalist rebels

Activities

1 Work in a group of four. Two people are to look at the text on page 14; and two people are to look at page 15. Each pair will need an outline map of Spain. You need to draw this yourself using the map on page 9 to help you. Using the information in the paragraph under your heading, label the map to explain how Spain developed.

2 Using both maps, make a display to explain why Spain today is not one nation, but several.

Is Basque terrorism Spain's biggest problem?

The Basque question – Spain's pressing problem

From a car bomb that killed four people in Madrid last October, an attack claimed by the separatist group ETA, to last November's 900,000 strong peace march through Barcelona to protest ETA's assassination of former health minister Ernest Llunch, the problem of Basque violence is ever present in Spanish society.

Polls show that about three-quarters of Spaniards believe Basque terrorism to be Spain's biggest problem.

'We are not going to allow them to impose terror on our country,' insists the Spanish Prime Minister José María Aznar.

The key issues

The two central issues in the separatist dispute are these: Who is a Basque? and What constitutes Basque territory?

Spain officially recognises three provinces as 'the Basque country'. Separatists, however, want another Spanish province, Navarra, to be included, as well as part of Southern France, to create a homeland for 3 million Basques.

Several world leaders and human rights organisations have called on ETA, blamed for about 800 deaths since 1968, to stop the killing. But the bombs and violence have continued.

Basque identity

Modern Basque nationalism sprang up a century ago as immigrants came into the region in search of factory jobs. The violent faction ETA started killing 32 years ago during the regime of General Franco, which ruthlessly suppressed the Basques and their language.

Much has changed since Franco's death in 1975. The Basque region now enjoys a broad degree of autonomy. As well as Basque-run schools, there is a Basque parliament and a Basque police force.

None of this, however, is enough for ETA and the other Basque nationalist political parties, with the result that the region's streets continue to simmer with tension, fear and a lot of anger.

Hopes for peace grew during a ceasefire called by ETA in September 1998. They were shattered a year ago, however, by a car bomb in Madrid, with both the government and ETA blaming each other for wrecking the ceasefire. For some, however, the struggle is already over. At one cemetery in San Sebastián, people killed by ETA are buried not far from a member of ETA itself. In life they stood at opposite poles. Now they are simply Basques.

△ **Figure 1.18** CNN report on the Basque question, 2001

△ **Figure 1.19** Car bombs have been part of ETA's fight for Basque independence

△ **Figure 1.20** The Basque landscape

▽ **Figure 1.21** Demonstrations against ETA violence

Foundation

1 Read the Internet article from page 16. Choose one of the photos opposite. Spend a few minutes looking at it. Copy and use the Learning from an Image frame (Figure 1.2) on page 6 to explain what you understand about the conflict in the Basque region.

Target

2 Look back at page 13. Use the Sense of Belonging diagram (Figure 1.15). Try to imagine how a Basque nationalist might feel about their place. Using the information from the Internet article on page 16 and the history on pages 14 and 15 try to complete the frame from the view of a Basque nationalist.

Extension

3 The conflict in the Basque region has to do with people's strong association with place. To help us to develop our understanding of this complex place we need to think critically about the information that we have. Look back at the Sense of Place frame on page 7. Use the frame to help you write about what you now know about the Basque region and the reasons for conflict there.

One Nation?

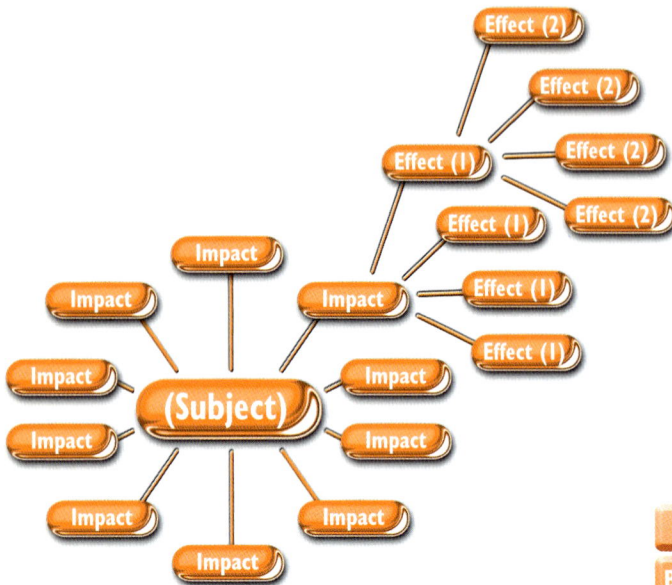

△ **Figure 1.22** Effects Wheel

Five questions to ask

1. IMAGES – What do you, and others, think this conflict is really about?

2. BACKGROUND – What has actually happened so far, and why?

3. SOLUTIONS – What are all the possible solutions that you can think of?

4. CHOICES – What are the best possible solutions?

5. ACTION – What, if anything, can you do about this issue?

IMAGES
BACKGROUND → CONFLICT → SOLUTIONS → CHOICES → ACTION

△ **Figure 1.23** Five Questions to Ask frame

Activities

4 You need to work in a group of four. You will be asked to look at the Effects Wheel (Figure 1.22). You then need to work with the others in your group on a copy of the Effects Wheel. If the Basque region were to gain independence, what could this mean? On your wheel note down as many impacts of this as you can.

5 From the list of impacts choose one and think about as many effects of this as you can. Make a note of these on your diagram.

6 You now need to form a new group of four; two of you need to join with another pair. Find out what is similar and what is different between the two groups' Effects Wheels.

7 In your new groups, use your Effects Wheel to explain how power and politics have altered the geography of the Basque region. Use the Five Questions to Ask frame above (Figure 1.23) to present your findings to the rest of the class. You may wish to do this orally, by writing, as a poster or by using PowerPoint.

ICT Activities

In Spain, there are two other regions where some of the people feel that they would be happier if they were independent from Spain. These regions are Catalonia and Galicia. The maps opposite show you where they are.

1 Working in a group of two or three, use the search facility on the Internet to find out as much as you can about the views in one of these regions.

2 Once you have gathered the information, use the Sense of Place frame on page 7 to help you to think critically about what you have learnt.

3 Then use the Five Questions to Ask frame to present your group's work. Decide with your teacher how you are going to present your work. Try to use a different style from Question 4 (Activities).

◁ **Figure 1.24** The regions of Spain

▽ **Figure 1.25** Galicia

△ **Figure 1.26** Catalonia

▷ **Figure 1.27** Basque country

Assessment tasks

A Spanish person's view

'My **ancestors** come from one of the most influential parts of the country in terms of independence – Galicia – but I think that Spain as an independent country should be politically, economically and industrially unified to participate fully in the European Union. I believe that all the citizens from the country should be able to recognise themselves as Spaniards and work together for the same goals to improve the way the country grows and develops. Not everyone agrees with me. Some areas have developed an identity different from the rest of the country. The differences are in language habits and the way that people have a strong identification with place. The three areas where this happens are: Catalonia, Basque country and Galicia.

Although these three regions hold different characteristics, I think that people should have a loyalty to Spain. Our language and nationality should be the major reasons for Spanish citizens to unite. Catalonia, the Basque country, and Galicia will not be able to keep their strength independently. The economic power that Catalonia has needs to be supported by the agricultural and military resources that other locations can offer. The same is true of the Basque country and Galicia. 'The Union makes the power' is a popular Spanish saying that people should apply in real life.'

△ **Figure 1.29** Working together?

△ **Figure 1.28** Adapted from an extract from a Spanish person's own website. The full text can be found at http://dons.usfca.edu/traned2/spain.htm

Target task

1 Read the Spanish person's view (Figure 1.28). Is this Spain? Then look at the cartoon (Figure 1.29). Can you explain how this person wants Spain to develop in the future and why? Use the How? and Why? frame below to help you note down the key points.

HOW?

WHY?

△ **Figure 1.30** How? and Why? frame

2 Either draw or stick a copy of the cartoon in the middle of an A4 sheet. Using your information from the How? and Why? frame, label the cartoon to show how it might relate to the development of a unified Spain.

3 Now look back on all the information about regional issues in Spain; you may wish to start on page 16. Use the Writing frame below (Figure 1.31) to help you explain the different view points.

The issue that we are thinking about is the Basque region in Spain.

Some people think that . . . .
because . . . .
They argue that . . . .
because . . . .
On the other hand . . . . .disagree with the idea that . . . .
They claim that . . . .
They also say that . . . .
My opinion is . . . .
because . . . .

△ **Figure 1.31** Writing frame

4 What do you thing should happen to the three regions: Basque, Catalonia and Galicia? Give reasons for your answer.

Extension task

A foreigner's view

'When I first went to Spain, in the early 1950s, the roads were very bad, accommodation sparse, comforts minimal. The sun shone, however, and it was not long before people from colder European countries such as the UK and Germany began to flock to Spain. Many Spaniards were uneasy about this and referred to tourism as *puteria*, or prostitution. The benefits were that tourism gave Spain an international dimension. Tourists, however, tended to keep to the beaten track, and much of Spain remained in rural isolation. From the 1950s on, I lived for a part of every year in a series of Spanish villages. It is in village life that Spanish qualities most reveal themselves. Conversations took as long as they had to, work was done in the course of time, and sometimes sitting down to a long lunch, I could feel the world outside withdraw, giving way to the small rich world of the table and the conversation. From my first visit on, simply being in Spain has always brought me a sudden joy, a physical tingle, from the light, from the landscape, from the language…'

△ **Figure 1.32** Adapted from *Insight Guide to Spain*

1 This visitor to Spain had one particular view of Spain, coloured by her experiences. Use the Sense of Place frame on page 7 to explain how you think that she developed this particular view of the place.

2 Review the regional information about Spain, which starts on page 16. When you have looked through it, you can use the Fact, Fiction or Opinion frame to help you organise the key pieces of information that you think are important. When you have selected the main statements, place them on the frame. It is best to start with 5–10 statements.

3 Use the information from the frame to explain to the foreigner in Figure 1.32 why their view of Spain might be a little simplistic and even romantic.

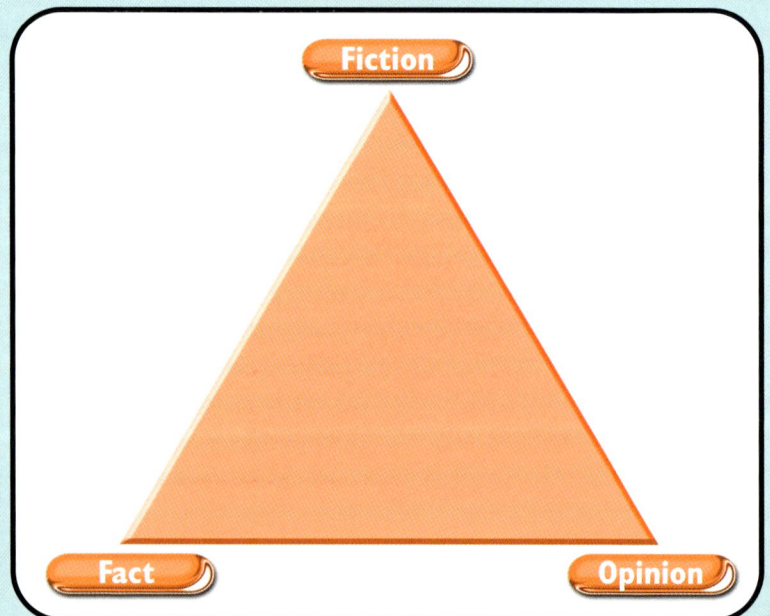

△ **Figure 1.33** Fact, Fiction or Opinion frame

Review

Activities

1 Look back at the images on pages 4 and 5. All of these are photos of Spain. These photos are very varied, as is Spain itself. The Spanish call Spain, Las Españas. Make a list of seven different photos or images that you would now choose to show someone to give them a glimpse of Spain. Explain why you would choose each photo, and what you hope that someone would think or feel about Spain from that image. You may want to search books or the Internet for some ideas.

2 Look back at your work on Spain. Make a note of all the ways that you have learnt about the country. Have a look at the Sense of Place frame on page 7. Look at Stage C. Describe how your knowledge of Spain has changed by doing this work.

3 Look at the quote opposite (Figure 1.34). How would you reply to the question, 'What is the real Spain?'

4 Think about another country that you would like to investigate. What sources of information would you need to build up a complex understanding of that place? Perhaps you might like to do this for yourself, deepening your own sense of place.

What is the 'Real Spain'? A gypsy caravan towed along an autopista [motorway] behind a gleaming BMW? Pale-skinned foreigners stretched out in neat rows on some sandy beach, broiling themselves a deep shade of lobster? The 'Real Spain' is an elusive concept.

△ **Figure 1.34** Adapted from *Culture Shock: A Guide to Customs and Etiquette, Spain* by Marie Louis Graff

△ **Figure 1.35** What is the real Spain?

2 Development: Global scale

What can maps show?

In this chapter you are going to investigate issues relating to development at the **global** scale. This will involve exploring world **patterns** and **distributions**. However, to know what these patterns mean in reality, development data needs to be investigated at a personal, local, regional, national and international scale. You will need to move between these scales to develop your own understanding of development issues. So, in a way, this chapter will take you through your own personal journey!

△ **Figure 2.1** The Earth from space

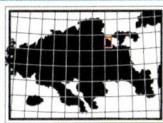

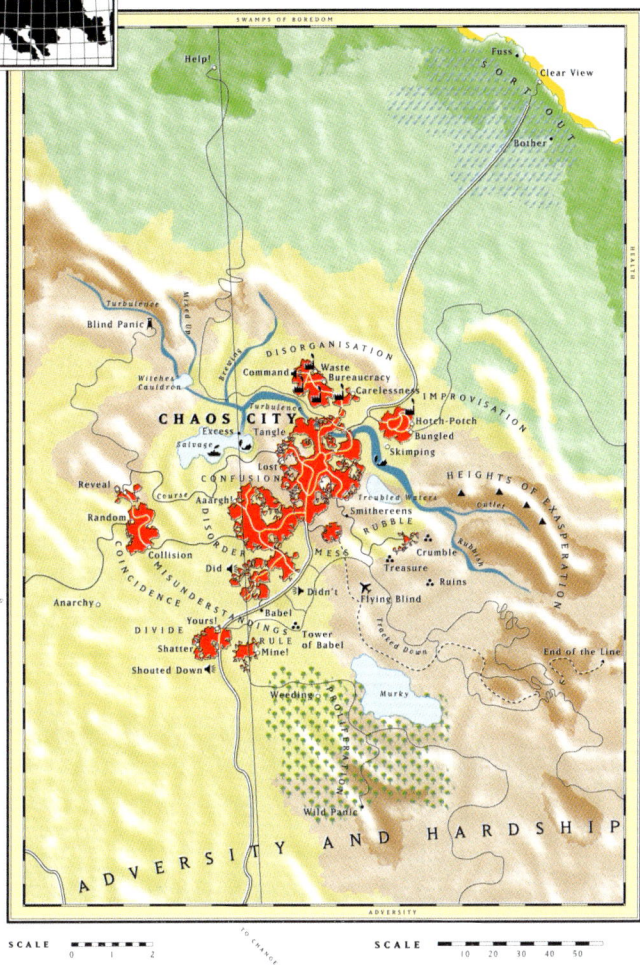

△ **Figure 2.2a** Map of Chaos from The Atlas of Experience

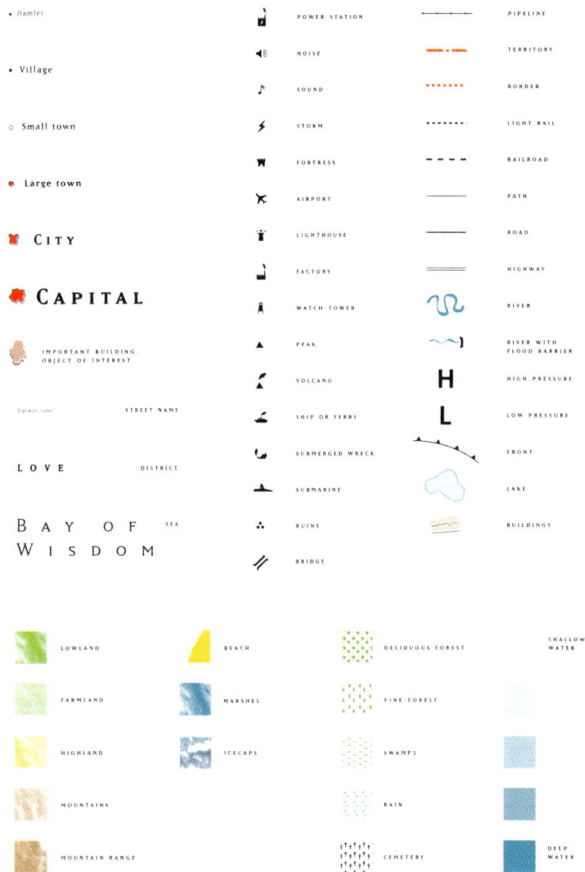

△ **Figure 2.2b** Key for Map of Chaos

Activities

1 Look carefully at the map above. It is an unusual map. The map comes from *The Atlas of Experience* produced by two Dutch **cartographers**, Louise Van Swaaij and Jean Klare. They suggest, 'An atlas never just shows you where you are, where you want to go to and how to get there. It also fires the imagination. Maps which chart rivers, mountains, towns, countries, far-away regions, oceans and continents can arouse intense feelings. An atlas combines reality and fantasy… At first sight the maps look like ordinary representations of far away places. On closer inspection, however, you will realize that you are surveying our shared world of thoughts and emotions.'

2 Note down your reactions to the map of chaos. Is it geography? What do you think/feel about the map?

3 Share your reactions with another person. Are there any similarities, differences in your reactions?

4 Using the legend (key) above, draw and colour your own map of experience. Think of things that you have choices about. Try to map them, and choose a title for your map.

5 When you have completed your maps, you might like to display them around the classroom. You will need to use your maps for work that you do elsewhere in this chapter, so spend some time looking at the maps that your class have produced and discuss what you think that they tell you about your personal choices and those of others in your class.

Do maps tell the truth?

The word 'geography', literally translated, means 'writing about the earth'. The map that you produced to show what choices you have is a way of writing about yourself. Often because something is mapped, we think that it is true. If we are to think critically about development at the global scale we are going to be looking critically at a number of world maps. We must be aware that these maps are not always exactly the same as reality. Some maps are selective in the information that they give, and represent somebody's particular view of the world. Look carefully at Figures 2.3, 2.4, and 2.5. All are maps illustrating aspects of global development.

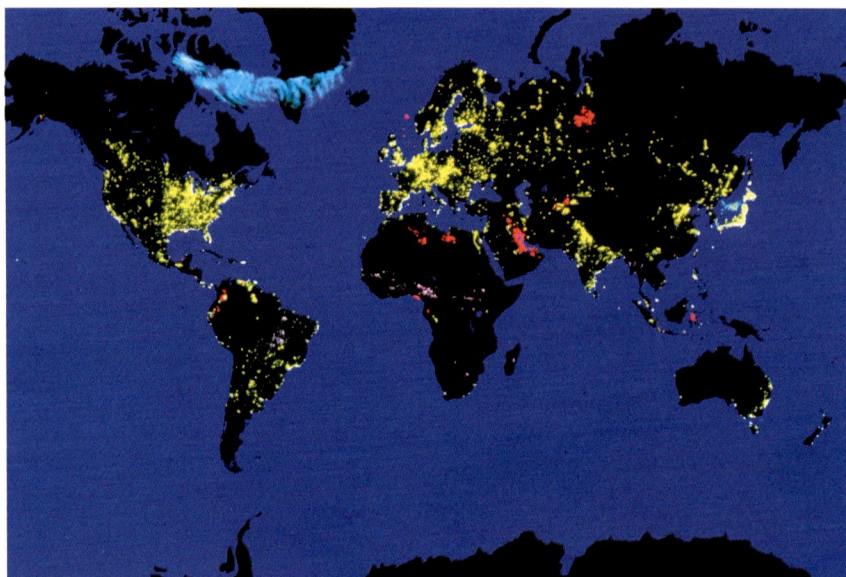

◁ **Figure 2.3** The Earth at night. This is based on data obtained from satellites of the American Defense Meteorological Program (DMSP). The yellow lights show street and building lights in urban areas. Oil production flares are shown in red and burning vegetation in pale purple. The green is light from squid-fishing fleets. The aurora borealis is the light blue band of colour

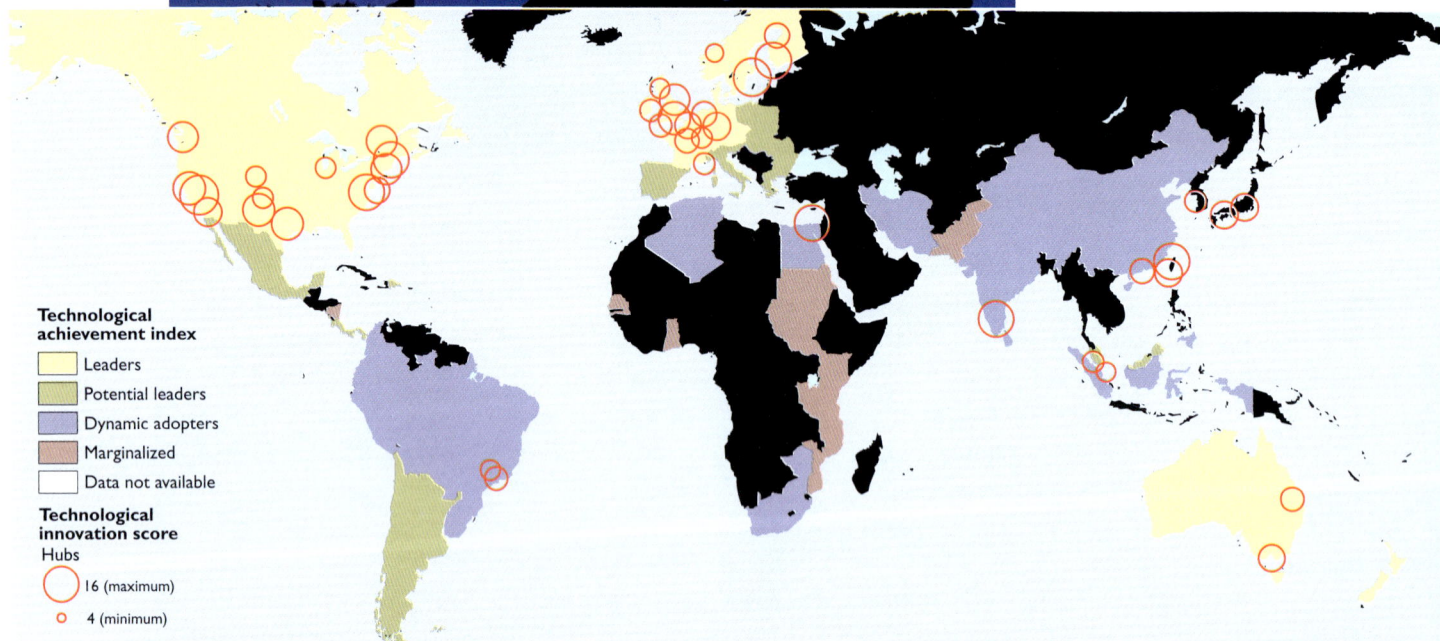

Technological achievement index
- Leaders
- Potential leaders
- Dynamic adopters
- Marginalized
- Data not available

Technological innovation score

Hubs
- 16 (maximum)
- 4 (minimum)

△ **Figure 2.4** The Geography of Technology map. In 2000, *Wired* magazine consulted governments, industry and the media to find the locations that are the most important in the new digital geography. Forty-six locations were found to be key locations, or hubs. They were given a score from 4–16, depending on how good they were at training, researching and several other factors. As well as these hubs, the map shows the countries that are leaders and potential leaders in technology. It also shows the countries that do not lead the way with inventions but are able to use and adopt the new technology. These are labelled as 'dynamic adopters'. There is also a group of countries on the map that is aware of the technology available but either due to a lack of money or skilled people have not been able to take advantage of what the technology can offer. These countries are labelled as 'marginalised'

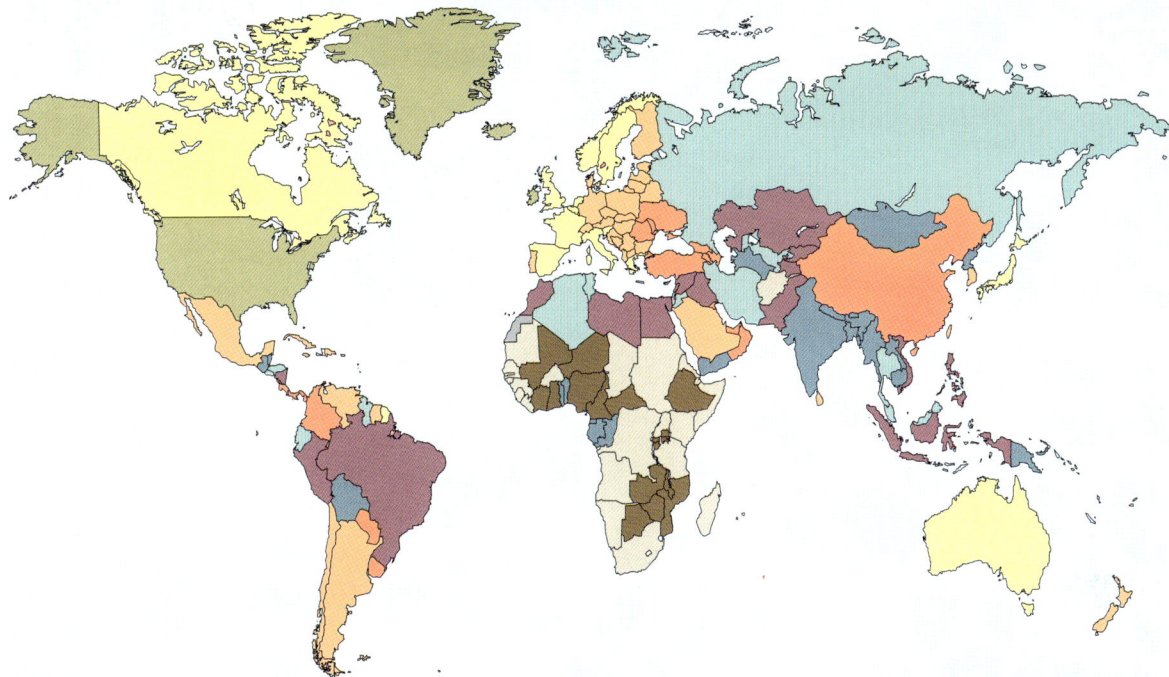

High expectancy

- 71.0 – 74.5
- 67.0 – 70.9
- 64.0 – 66.9
- 62.0 – 63.9
- 60.0 – 61.9
- 55.0 – 59.9
- 45.0 – 54.9
- 35.0 – 44.9
- 25.9 – 34.9

Low expectancy
- No data

Measure: Disability adjusted life expectancy at birth, both sexes, estimates for 1997

△ **Figure 2.5** Healthy Life Expectancy from the World Health Report, 2000

Activities

All three maps are representations of the Earth. All are very different. You need to work with a partner, and together you will be asked to look in detail at one of the maps (Figures 2.3, 2.4 or 2.5). You will also need to use an atlas with a world map showing country outlines and names. Work with your partner to decide.

1 What do you like and/or dislike about the map?

2 Is it easy or difficult to work out what the map is showing you? How could the layout of the map be improved?

3 Write four sentences to describe what you think that your map is telling you about development.

4 Using your atlas, choose four countries from four different continents that you want to find out more about. List the countries and note down three things that are true for each country that you can find out from your map.

5 Do you think that your map is representing the truth?

6 How would you describe the global pattern shown in your map. Use at least four sentences to describe the pattern.

You need to make certain that both you and your partner have noted down your answers, since in the next activities you will become the expert for that map.

Whose choice?

The Human Development Report of 2001 stated that 'Human development is about much more than the rise or fall of national incomes. It is about creating an environment in which people can develop their full **potential** and lead productive, creative lives in accord with their needs and interests. People are the real wealth of nations. Development is thus about expanding the choices people have to lead lives that they value.'

Activities

You need to work in a group of three, each of you will have worked on a different map for the Activities on page 27.

1 Spend some time talking to each other about your maps. Use your answers to question 1 on page 27 to help you describe what your map shows and explain your reasons for your observations.

2 Look at the cartoon (Figure 2.6). What is your group's reaction to it? What do you think it is saying about development at the global scale?

3 Make your own notes about what you learn from the others in your group.

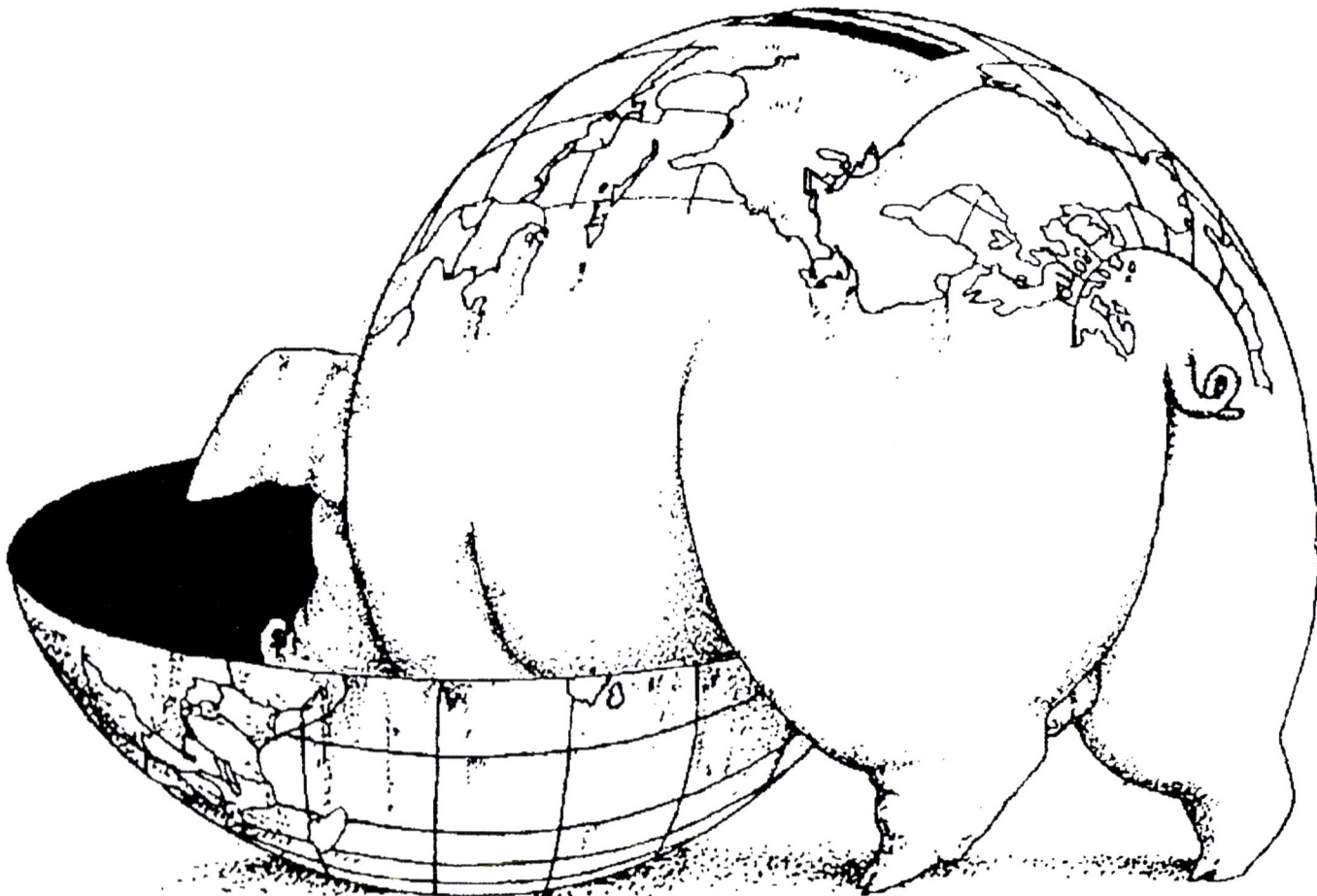

△ **Figure 2.6** Global connections?

Development patterns

Foundation

4 Look back on Figures 2.3, 2.4 and 2.5. You may wish to use your notes from the Activities on page 27 to help you. Use this information to label a blank world map outline with what you know about global development.

Target

5 You will need six colours, and a blank world map outline. Label the world map to show world development patterns. Try to classify your labels to say whether they show information at the:

- **personal scale** – your view
- **local scale** – a small scale area anywhere in the world
- **regional scale** – part of a country

- **national scale** – a whole political unit, or country
- **international scale** – link between two or more countries
- **global scale** – whole world

6 Is there any comment that you want to add to your map?

Extension

7 Think about the work that you have just completed. Create your own world map to show global development patterns. Think carefully about how you did it, and how you developed your thinking. Now read the quote below (Figure 2.7).

8 Explain what you think that Kofi Annan means, and describe how you used dialogue to develop your thinking.

▽ **Figure 2.7** UN Secretary General, Kofi Annan, 2001

'I see dialogue as a chance for people of different cultures and traditions to get to know each other better, whether they live on opposite sides of the world, or on the same side of the street.'

Do you know that...

...emailing a 40-page document from Chile to Kenya costs less than 10 cents, faxing it about $10, and sending it by courier $50?

...in 2001 more information can be sent over a single cable in a second than in 1997 was sent over the entire Internet in a month?

...the cost of transmitting a trillion bits of information from Boston to Los Angeles has fallen from $150 000 in 1970 to 12 cents today?

... there are expected to be 1 billion users of the Internet in 2005?

...in two years from 1998 to 2000 Internet users increased from 1.7 million to 9.8 million in Brazil, from 3.8 million to 16.9 million in China and from 2 500 to 25 000 in Uganda?

...Cuba developed the only vaccine against meningitis B, through biotechnical research providing national immunisation by the late 1980s?

...India's information and communication technology exports rose from $150 million in 1990 to nearly $4 billion in 1999?

...between 1992 and 1997 Vietnam reduced the death toll from malaria by 97% and the number of cases by almost 60% by developing and using locally produced, high quality drugs?

...in Brazil a team of computer scientists, commissioned by the government, have designed a basic computer for around $300?

... pharmaceutical sales in Africa are forecast to be just 1.3% of the global market in 2002?

...just 0.1% of the 25 million in sub-Saharan Africa have access to HIV/AIDS drugs?

Note: You may want to use the Internet to find out the current dollar exchange rate.

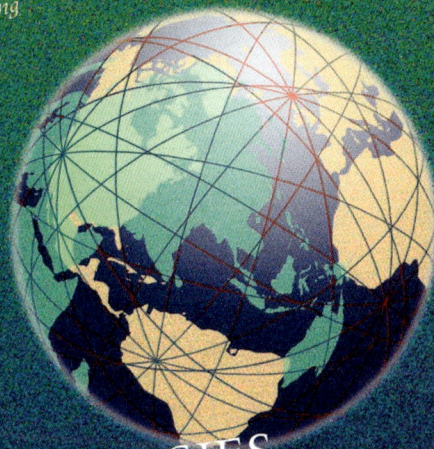

HUMAN
DEVELOPMENT
REPORT 2001

Technology networks are transforming
the traditional map of development,
expanding people's horizons
and creating the potential
to realize in a decade
progress that required
generations in the past

MAKING
NEW TECHNOLOGIES
WORK FOR
HUMAN DEVELOPMENT

△ **Figure 2.9** Every year since 1990, the United Nations Development Programme has commissioned the Human Development Report (www.undp.org/hdro) by an independent team of experts to explore major issues of global concern. The report looks beyond income as a measure of human progress

Activities

1 Look at the information. Choose five of the statements that surprised you the most. Now rank your statements 1 to 5; 1 will be the most surprising.

2 Share your five statements with three other people in your class. Do any of the statements appear in all of your lists? Can you explain any similarities and differences between your lists.

3 On a blank world map outline, label the countries that the speech bubbles refer to. Try to add labels to your map to show as much information as you can without crowding your map. Think about your map design. Make sure that your map has a title, key and scale.

4 Write a paragraph to describe what you think that your map illustrates.

5 Using your map and paragraph, predict what you think that the 2001 Human Development Report might say about technology and development.

What is the role of technology in development?

The Human Development Report 2001, commissioned by the United Nations, argues that new technologies can play a huge role in reducing world poverty; it does not hold the view that technology is a luxury for people in rich countries.

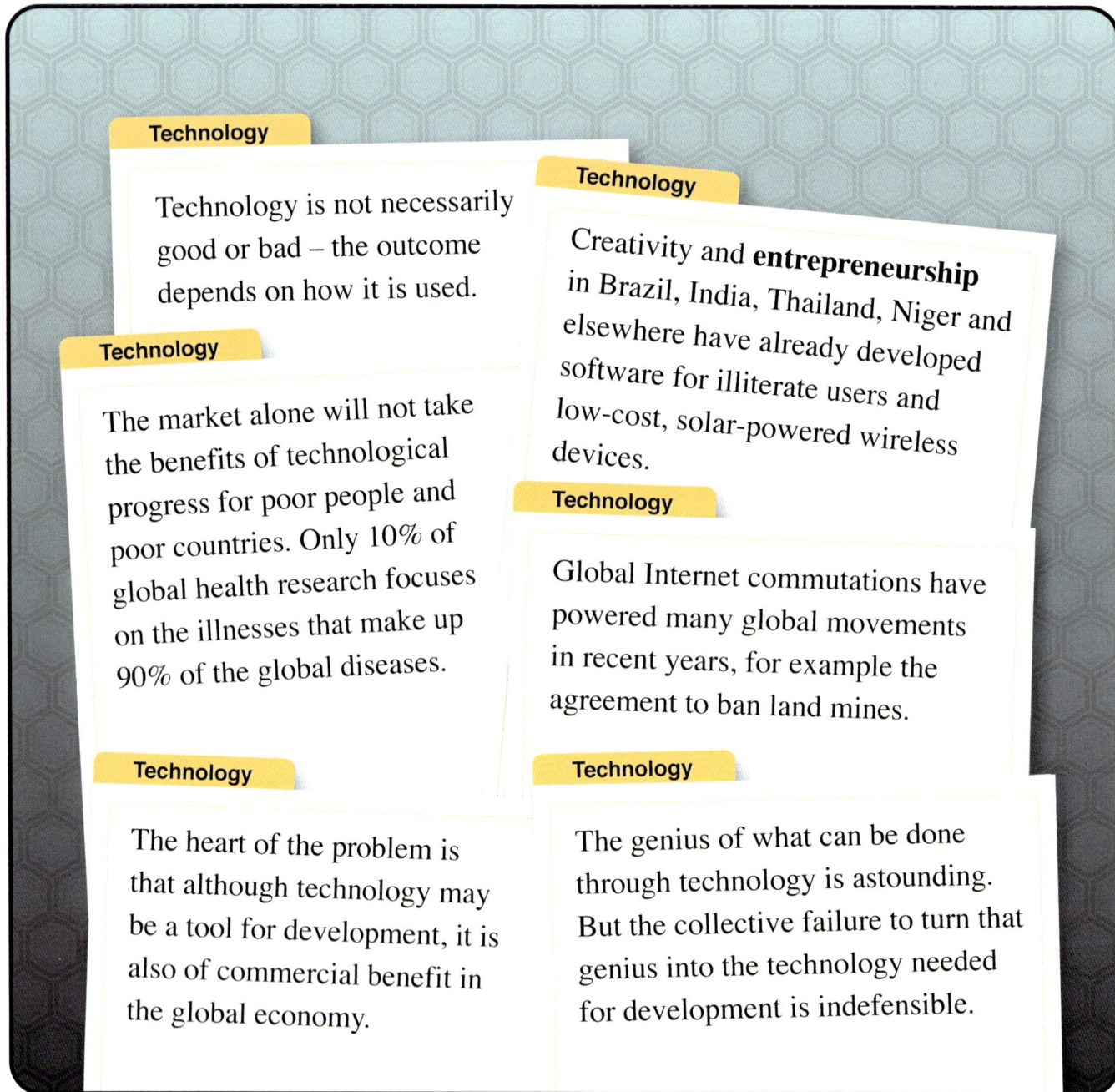

Technology

Technology is not necessarily good or bad – the outcome depends on how it is used.

Technology

The market alone will not take the benefits of technological progress for poor people and poor countries. Only 10% of global health research focuses on the illnesses that make up 90% of the global diseases.

Technology

The heart of the problem is that although technology may be a tool for development, it is also of commercial benefit in the global economy.

Technology

Creativity and **entrepreneurship** in Brazil, India, Thailand, Niger and elsewhere have already developed software for illiterate users and low-cost, solar-powered wireless devices.

Technology

Global Internet commutations have powered many global movements in recent years, for example the agreement to ban land mines.

Technology

The genius of what can be done through technology is astounding. But the collective failure to turn that genius into the technology needed for development is indefensible.

△ **Figure 2.10** Quotes from Human Development Report, 2001

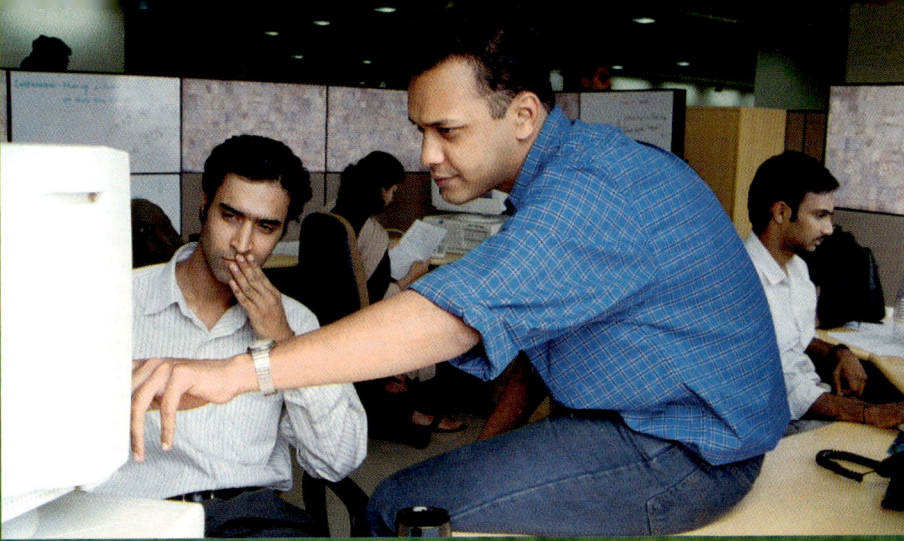

△ **Figure 2.11** Computer training in India – technological change dramatically raises the value every country should place on investing in the education and training of its people

△ **Figure 2.12** Kosovan refugees surf the web – the Internet is breaking barriers of geography, making it possible for remote communities to benefit from non-formal education and life-long learning as well as access to information about local services, recreation and other programmes

Case Study 1

Open source software is the outcome of many voluntary contributions from around the world. The details of how the software cannot be hidden, but must be kept open for all to see. It is low cost, often free, allowing governments to make their information and communication technology budgets go further.

Open source software could speed the ICT revolution if its uses take off in a wide scale.

Case Study 2

The Koyhmale Community Radio in Sri Lanka uses radio as a gateway to the Internet for its listeners in remote rural communities. Children or their teachers send requests for information about school topics. The broadcasters search for information on the Internet, download it and make it available by making a broadcast around the information, mailing it to the school or placing it in the radio station's open access resource centre. This is made available in local languages rather than English.

△ Case studies, Human Development Report, 2001

◁ **Figure 2.13** A satellite dish in Afghanistan – new technologies need not pass developing countries by. Making satellite dishes domestically means lower prices, wider access to broadcasts and more job opportunities

What's your view?

Activities

1 Use the information on pages 32–3 to make an oral presentation about Development and Technology. The presentation will be in the form of a 'SOFT' report: Strengths, Opportunities, Failures, and Threats to technology-aiding development.

2 The class to be divided into two halves: A and B.

3 Each half will then work in smaller groups of three or four.

4 Your group will have to work on one paragraph of the SOFT report. Use a copy of the frame below to organise your group's thoughts.

Talk title	Either A or B's SOFT report on the HDR 2001	
Paragraph number	1 = Strengths 3 = Failures	2 = Opportunities 4 = Threats
Main themes		
Key words to use		
Big point to begin paragraph		
Sentences which offer detailed evidence relating to the big point	1 2 3 4	

△ **Figure 2.14** Oral presentation Thinking frame

5 Everyone in your group will need their own rough copy of this Thinking frame.

6 Now decide who in your group is going to take on each of these tasks:
- One person will need to give the speech.
- One person will need to represent your group with the others from your half of the class – a collaborator.
- One person will go to the other half of the class to observe the other group working on the same paragraph as yourselves – observer.
- One person will become an evaluator.

Can technology make a difference?

7 Each person now has a job to do. Your teacher may limit the amount of time that you have to do these jobs:
- The speech needs to be written.
- The collaborators need to meet to make sure that each paragraph joins up with the others.
- The observers need to go and see if they can benefit from the other group's thinking.
- The evaluators need to meet in two groups, A and B. They need to decide up to five criteria on which they will judge the two presentations.

8 Go back to your group. Share the information that you now have. Work together to write your final paragraph.

9 When you are ready to make the presentations, the speakers for each half of the class will need to sit together, as will the evaluators. The observers and collaborators will form the audience.

10 When the presentations have been made, the evaluators will need to share their judgements and give their reasons and criteria.

11 When you have completed this task, take a few moments to think about how you worked as an individual, as a member of a small group and as a member of a larger group. Are there any messages from the way that you worked that might be important when thinking about technology for global development?

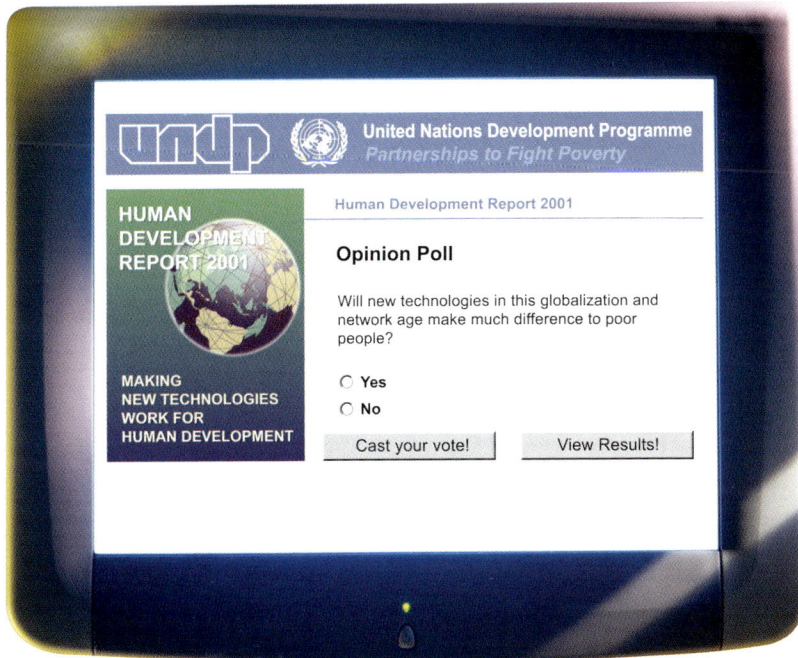

△ **Figure 2.15** UNDP Opinion Poll

ICT activity

After the publication of the report, the UNDP held a poll on their website to see if people thought that new technologies could make a difference to the poor. If you were to respond, how would you vote and why? Plan or construct a PowerPoint presentation to explain how and why you made your decision to vote yes or no. The full report can be found at www.undp.org/hdr2001/
The poll was found at http://roo.undp.org/hdr/poll.cfm

Is technology the answer?

The Human Development Report 2001 argues that technology is part of the solution to aiding development at the global scale. Not all countries need to be at the cutting edge of research and development, but all countries need to be linked to the network age, so that they can adopt and use the best technology. The report goes on to say that 'A unique advantage of using technology for human development is that there is no need to retread the same gradual path once followed by developed countries.'

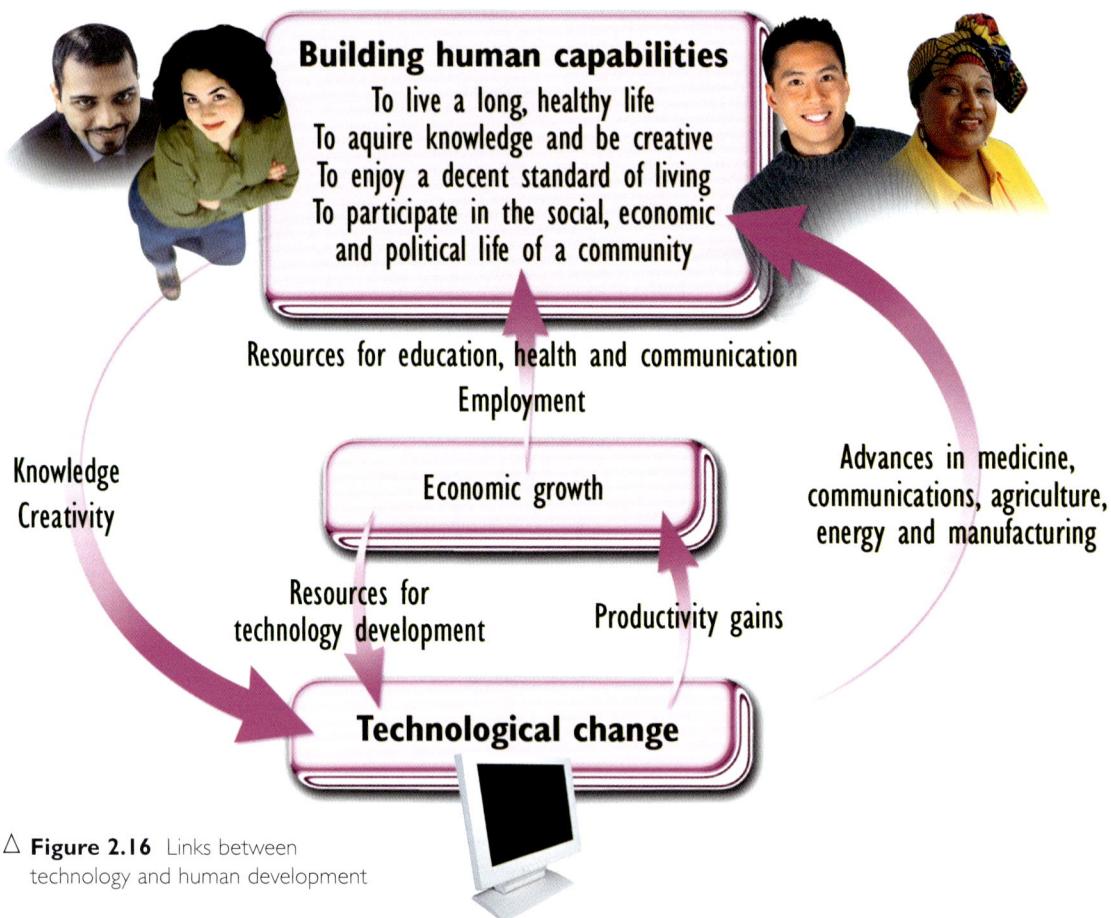

Building human capabilities
To live a long, healthy life
To aquire knowledge and be creative
To enjoy a decent standard of living
To participate in the social, economic and political life of a community

Resources for education, health and communication
Employment

Knowledge
Creativity

Economic growth

Advances in medicine, communications, agriculture, energy and manufacturing

Resources for technology development

Productivity gains

Technological change

△ **Figure 2.16** Links between technology and human development

The World Wide Web is too expensive for millions of people in developing countries, partly because of the cost of computers. In January 2001 the cheapest Pentium III computer was $700–hardly affordable for low-income communities. The text-based pages of the Internet puts it out of reach for illiterate people.

To overcome these barriers, academics at the Indian Institute of Science and engineers at Bangalore designed a handheld Internet appliance for less than $200. It will provide Internet and email access in local languages, with touch screen functions. Future versions promise speech recognition and text to speech software for illiterate users.

△ **Figure 2.17** Breaking barriers to Internet access (adapted from *PC World 2000*; Simputer Trust 2000; Kirkman 2001)

◁ **Figure 2.18** Solar power used to heat a kettle, Zanzibar, Tanzania

Activities

Foundation

1 Look carefully at the diagram to show how technology can aid development (Figure 2.16). Think for a moment about yourself and the choices that you make. You may wish to refer back to your experience map from the activity on page 25. Write a few sentences to describe how technology has helped you to develop. Here are some sentence starters:

Technology has helped me because … Also … Another thing is … Finally …

2 Look at the photo for Zanzibar, Tanzania (Figure 2.18). Redraw Figure 2.16 using the information from the case study to show how technology will potentially aid development in Zanzibar.

Target

3 Re-read the information on page 28 about what development means. Think about being a young person, the age that you are now, living in Zanzibar. Now complete the frame below:

Young people's choice	Kaanwa with electricity in 5 years	Kaanwa without electricity in 5 years
Local environment		
Local community		
Ability to earn money		
Ability to take part in decision-making		
Anything you want to add		

Extension

4 Read the text 'Breaking barriers to Internet access'. Redraw Figure 2.17 to show how technology could aid development in Bangalore.

5 How might the future be different if this innovation had been developed by a commercial company?

Most countries not on track to meet UN's 2015 goals

Human Development Report data indicates need for new initiatives

Mexico City, 10 July 2001–Last September at the United Nations Millennium Summit, world leaders agreed on a set of goals for development and poverty eradication to achieve by the year 2015.[1] But, according to new analysis in the **Human Development Report**, many countries are not on track to achieve these goals.

- Eleven million children below age five still die every year from preventable causes–about 30,000 a day.

- Nearly one billion people still need access to safe drinking water.

- There are still 1.2 billion people who live on less than $1 a day.

[1] For the text of the Millennium Declaration, see www.un.org/millenium/declaration/ares552e.htm

Does aid have a role?

Despite the huge increase in the use of the Internet in developing countries, the **digital divide** still exists. Much of the technology is concentrated in wealthy countries and even old inventions like grid electricity is out of reach for 2 billion people. Internet costs are particularly high in developing countries. Monthly Internet access charges are 1.2% of average monthly income for a typical US user, compared with 191% in Bangladesh (2001).

In 2000, at the G8 Summit, protesters mocked international efforts to use technology to meet the needs of the poor. 'We can't eat computers' complained the leader of a group campaigning for debt relief. 'People are dying.' To make their point, members of the group set fire to a laptop computer. Within development circles some have worried that the emphasis on technology might distract donors and take money away from more traditional development targets.

Activities

1 Read the press release (Figure 2.19). How could the protesters at the G8 summit use this information to support their view?

2 Look at the information in the table below. Quite often these numbers are so large that it is difficult to think about what they really mean. Consider for a moment if it were you or a member of your family experiencing the deprivation. You may like to look back at your experience map from the Activity on page 25. How would your choices be different if you were suffering these deprivations?

3 What would you say to the protesters at the G8 summit? Do you agree or disagree with them? Why?

Serious deprivations in many aspects of life

Developing countries

Health

- 968 million people without access to improved water sources (1998)
- 2.4 billion people without access to basic sanitation (1998)
- 34 million people living with HIV/AIDS (end of 2000)
- 2.2 million people dying annually from indoor air pollution (1996)

Education

- 854 million illiterate adults, 543 million of them women (2000)
- 325 million children out of school at the primary and secondary levels, 183 million of them girls (2000)

Income poverty

- 1.2 billion people living on less than $1 a day, 2.8 billion on less than $2 a day (1998)

Children

- 163 million underweight children under age five (1998)
- 11 million children under five dying annually from preventable causes (1998)

Developed countries

- 15% of adults lacking functional literacy skills (1994–98)
- 130 million people in income poverty (1999)
- 8 million undernourished people (1996–98)
- 1.5 million people living with HIV/AIDS (2000)

▷ **Figure 2.20**

Although this chapter has been concerned about global development and global patterns, we often need to work at a smaller scale to find out what these patterns really mean. In this assessment activity you are going to investigate the impact of global development patterns at the national scale. India is a country that has experienced both positive and negative effects of technology.

India is home to one of the global technology hubs – Bangalore (see Figure 2.4), yet India ranks 63rd in the technology achievement index. This is because of the big difference in technological achievements among Indian states. The country has the world's seventh largest number of scientists and engineers, and yet in 1999 the adult illiteracy rate was 44%.

The costs to India of providing a university education for 100 000 Indian professionals is $2 billion per year. Skill shortages in Europe, Japan and the US have led workers to move between countries. In 2000 the US approved more work visas for skilled professionals. Of the 81 000 visas approved between October 1999 and February 2000, 40% were for people from India and more than half were for computer-related occupations. This brain drain makes it very difficult for developing countries to keep the people they need for the development of technology.

Target tasks – India and technology

1 Read through the information on page 40. Make a list of all the ways that technology has had a positive effect on India.

2 Now re-read the information on page 40 and this time list all the negative effects that technology has had on India.

3 Write a paragraph to explain if you think that technology is helping India to develop as a nation. You may wish to use the 'I think…' frame opposite to help you. Try to include as much factual information as possible to support your view.

I think that technology is helping/not helping India by …
Also I think …
Not everyone would agree with me because …
Also …
However, in conclusion I think …

Extension tasks – the Global divide

Features of Technology Development 2001	
• Costs of Internet access vary; very cheap in some locations, hugely expensive in others	• Access to mobile phones is most difficult where they are most needed • •

Preferred features of Technology and Development	
• Development of open source software • Development of cheaper Internet access relative to income	• Access to technology hardware • •

△ **Figure 2.21** Features frame

4 Look carefully at the information on page 42. Refer back to the work that you have completed while investigating global patterns of development. Using the Features frame above, copy and add to it your ideas. You need to think about the global structures and features that were in place when the report was written. You then need to think about some of the structures and features that the report suggests need to be in place to support technology and development.

5 Look at your completed Features frame. Describe the global challenges that need to be faced if your preferred features are to become a reality.

6 Describe the problems that you anticipate, the things that you fear and describe also any benefits and things that you look forward to.

7 Look at the press release on page 38. The table highlights some of the goals that the UN agreed on for poverty and development eradication. What goals would you create to enable global developments to take place?

Assessment tasks

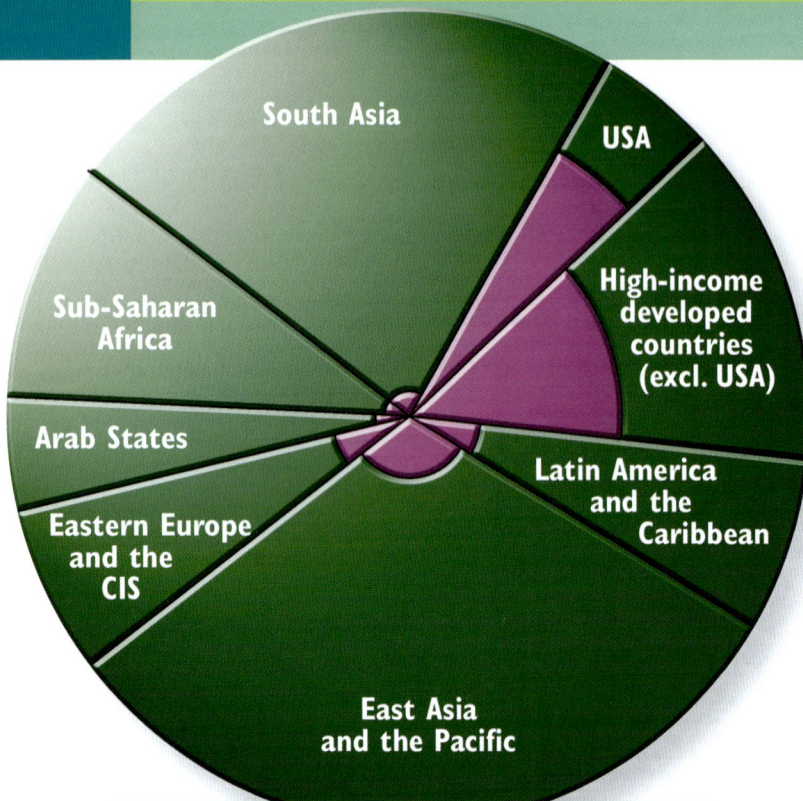

South Asia

USA

High-income developed countries (excl. USA)

Sub-Saharan Africa

Arab States

Eastern Europe and the CIS

Latin America and the Caribbean

East Asia and the Pacific

The large circle represents world population. Pie slices show regional shares of world population. Purple wedges show Internet users.

Internet users (as % of population)		
	1998	2000
USA	26.3	54.3
High-income developed countries (excl. USA)	6.9	28.2
Latin America and the Caribbean	0.8	3.2
East Asia and the Pacific	0.5	2.3
Eastern Europe and CIS	0.8	3.9
Arab States	0.2	0.6
Sub-Saharan Africa	0.1	0.4
South Asia	0.04	0.4
World	2.4	6.7

△ **Figure 2.22b** Internet users as a percentage of population, 2001. Source: See Fig. 2.22a

◁ **Figure 2.22a** Internet users, 2001. Source: Human Development Report Office based on data supplied by Nua Publish 2001 and UN 2001

More people have access . . .
Millions of Internet users

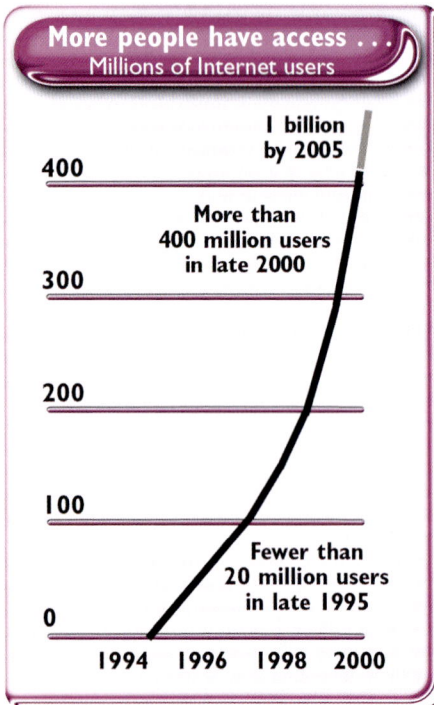

1 billion by 2005

More than 400 million users in late 2000

Fewer than 20 million users in late 1995

△ **Figure 2.23** Access to the Internet. Source: Nua Publish 2001

. . . to more information . . .
Number of Websites

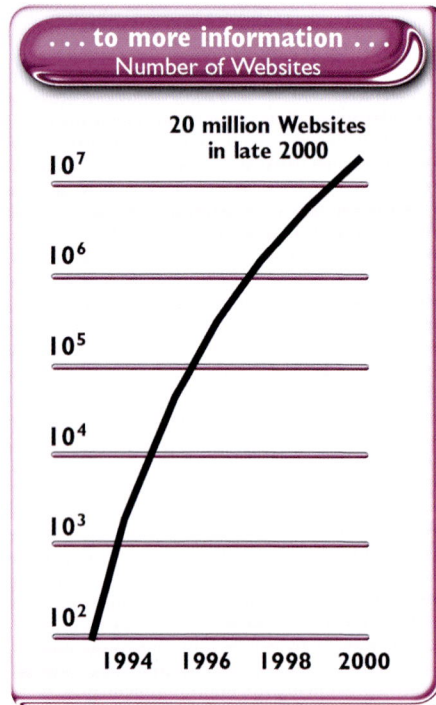

20 million Websites in late 2000

△ **Figure 2.24** Number of websites ($10^2 = 100$, $10^3 = 1\,000$, $10^4 = 1\,000\,000$ etc). Source: Robert Hobbes Zakon, 2000. Hobbes Internet Timeline

. . . at a lower cost
Transmission cost

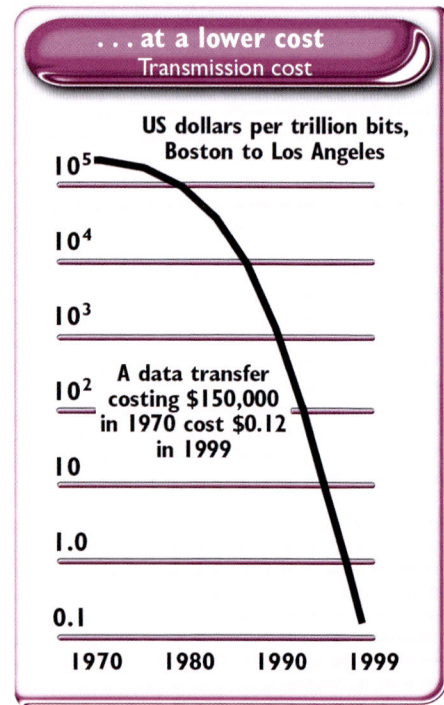

US dollars per trillion bits, Boston to Los Angeles

A data transfer costing $150,000 in 1970 cost $0.12 in 1999

△ **Figure 2.25** Costs of data transfer. Source: Cox and Alm, 1999. The New Paradigm

Activities

1 Look back at the experience map that you drew to represent your choices on page 25. Now draw a new experience map to show the choices that you have related to global patterns of development. What features do you think are important now, that you were not so aware of before?

△ **Figure 2.26** Development is complex

2 Have a careful look at the cartoon above. Talk to someone else about what you think the cartoon is saying. Now working on your own, explain what you think that the cartoonist was trying to communicate about development and aid. Why do the authors of the Human Development Report 2001 not think that aid alone will support global development?

3 The authors of the Human Development Report 2001 argue that to support development we need to change the way technological developments are shared. Try to devise a cartoon to show one of the changes that the report is calling for. You may wish to look back at pages 32 and 33 to help you.

4 Think about how you have learnt about global patterns of development from this book. Do you share knowledge or do you always compete? The Human Development Report 2001 talks about sharing technology as a way to improve choices. What does your experience tell you about both sharing knowledge and competing?

What is a natural hazard?

Earthquakes and volcanoes are both **natural hazards**. These are physical processes or events which have the potential to cause harm to life or property. While earthquakes and volcanoes happen in many places around the world every day, it is when they affect people that we call them hazards. Both earthquakes and volcanoes are caused by movements within the structure of the earth. The study of these movements is called Tectonics, so earthquakes and volcanoes are often described as '**tectonic activity**'.

Some countries have spent huge amounts of money trying to prevent the problems caused by natural hazards from affecting the population.

▽ **Figure 3.1**

'I felt my heart beating harder and faster than ever before.'

'The sight of such appalling death and destruction was unbearable.'

'As the sky darkened, a strong wind blew toward us. We knew something terrible was about to happen.'

'I was shaking uncontrollably, it had affected me more than I thought.'

'All I could hear were the screams of those around me.'

The quotes and the photos in the diagram are linked to the following topics:

- **Living in a war zone**
- **Being caught in a volcanic eruption**
- **Driving in a Formula 1 race**
- **Riding a theme park roller coaster**
- **The effects of an earthquake**

The following words can also be linked to the photos and the quotes:

Fear
Terror   Helpless   Memorable
Life Altering   Expensive
Powerful   Unfair   Short Term
Controlled   Predictable   Long Term

▽ **Figure 3.2** Living in tectonically active areas

Activities

1 Put the pictures and quotes together with the topic you think best explains them.

2 How many descriptive words can you link with each photo?

3 Which of these words make you think of negative things and which make you think positively?

4 If people know that living in tectonically active areas can be dangerous, why do they do it?

5 From the table below, write a list of the three best and the three worst things about living in these areas.

	Benefits	Problems
Environmental	• Fertile soils created by volcanic eruptions • Spectacular scenery is often created by tectonic activity	• Hazardous living conditions • Potential damage to ecosystem
Social	• Throughout history communities all over the world have grown up next to natural hazards like volcanoes. They have formed part of their religion and culture, creating many unique civilisations	• Damage to a community can take years to rebuild • Emotional damage to people who have suffered will never go or be repaired
Economical	• Tourist attraction (e.g. Vesuvius and Pompeii) • Precious metals and gem stones are found in areas of tectonic activity	• Industry can be destroyed by tectonic activity • Crops can be lost and livestock killed

What do you already know about earthquakes and volcanoes?

The results of volcanic eruptions and earthquakes have shaped the world we live in. In the UK this tectonic activity happened millions of years ago, but we still see the results in our landscape. The Highlands of Scotland, the English Lake District, and the mountains of Wales have all been formed from tectonic activity. However, there are areas of the world where many people live that are being changed by dramatic volcanic eruptions and earthquakes on a regular basis.

To help understand more about the lives of people in these areas, we must first explore how much we know about these types of event.

Facts and Opinions

- **Geography is made up of facts and opinions.**
- **People make decisions based on facts and opinions.**
- **Inaccurate facts and biased opinions can mislead decision makers.**

We can control the impacts of tectonic events to prevent loss of life.

In some countries a lot of money is spent each year to try and predict when earthquakes or volcanic eruptions are likely to happen.

As people have found out more about the structure of the earth, the reasons for earthquakes and volcanoes have become clearer.

The environments created by tectonic activity can have special benefits for people.

In some parts of the world people live with the threat of these natural hazards from day to day.

People have been living in hazardous environments since the beginning of life on earth.

Earthquakes and volcanoes create nothing but misery and suffering wherever they occur.

△ **Figure 3.3**

There are many ways of studying earthquakes and volcanoes. In geography we tend to split these into two main categories: Physical and Human.

Physical Factors

Research can be done on the structure, size and location of earthquakes and volcanoes.

◁ **Figure 3.4**
Ways in which volcanoes and earthquakes can be studied

Human Factors

Living with earthquakes and volcanoes often means adapting your way of life. Coping with death and destruction can alter the way a community's culture develops and how it does business with the rest of the world.

Activities

You can use these activities at the beginning of this chapter to show how much you already know about earthquakes and volcanoes and at the end to show how much you have learnt.

1 On a larger copy of the table, say which statements you think are facts and which are opinions. Explain where the evidence comes from for your choices.

2 Find out about a recent earthquake or volcano. Can you identify facts and opinions in the reports you read?

3 What is the danger of treating facts and opinions in the same way?

	Start of chapter		End of chapter	
	Fact or opinion?	How do you know?	Fact or opinion?	How do you know?
Statement 1				
Statement 2				
Statement 3				

fore we look at the effects of ectonic activity we need to investigate the causes. To do that we must start at the beginning with the structure of the earth.

As you can see from the diagram (Figure 3.5) the earth is not just a large solid object. It has quite a complex structure with most of the planet being a mixture of solid, semi-molten and liquid rock.

To explore more of the tectonic processes which affect us we have to look at the section of the earth's structure nearest the surface. This is shown in Figure 3.6. The surface of the earth is called the crust. It 'floats' on top of the **mantle**. The crust is formed when molten rock breaks through areas of weak crust and hardens to form new crust. This is called a constructive plate margin. New crust is formed and the ocean floor spreads out from this area. One of the largest areas of new crust formation is the North Atlantic Ridge (see map, Figure 3.8). Sometimes these areas are concentrated in '**hotspots**' and new crust is formed just in these places. Hawaii is formed at a 'hotspot'.

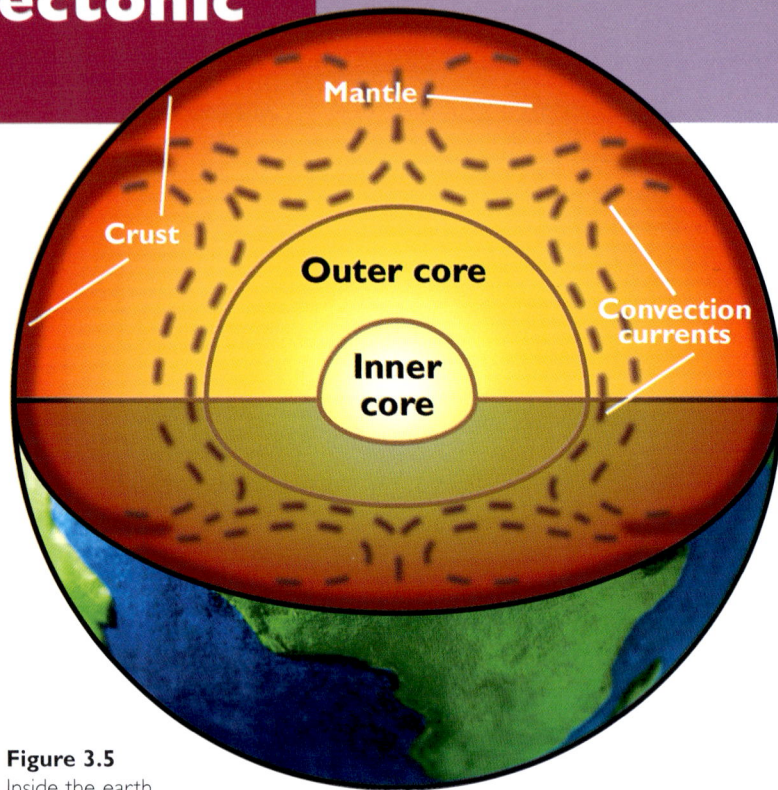

△ **Figure 3.5**
Inside the earth

Some parts of the crust are more dense than others, so when they meet one is forced to dip underneath the other. This sinking crust eventually gets reabsorbed into the mantle. These areas are called destructive plate margins (Figure 3.7) as some crust is destroyed by the process of **subduction**. It is at these destructive plate margins where much of the tectonic activity we are interested in occurs. With the friction of the two plates moving against each other, there is a great deal of heat created which can weaken the crust and lead to volcanic eruptions. This friction can also cause great pressures to build up, to the point where two plates may stick together, only to break apart in a violent movement when the pressure gets too much. This causes earthquakes.

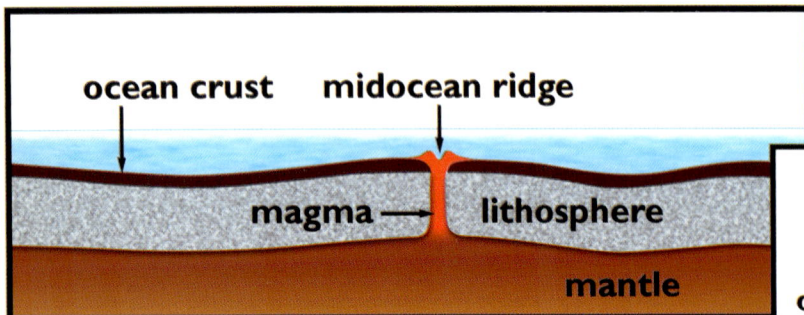

△ **Figure 3.6** Constructive plate margin or mid-ocean ridge

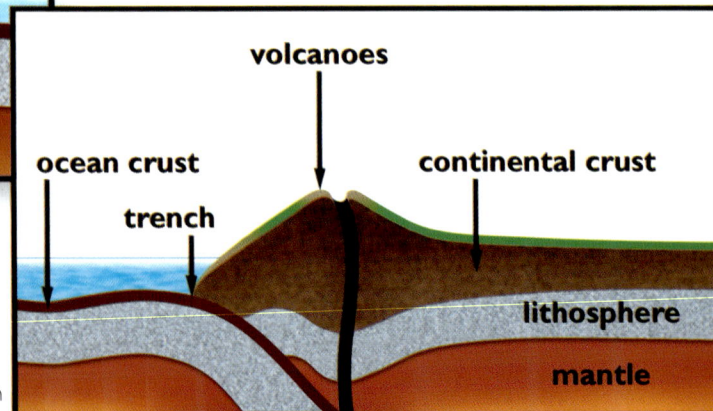

▷ **Figure 3.7** Destructive plate margin

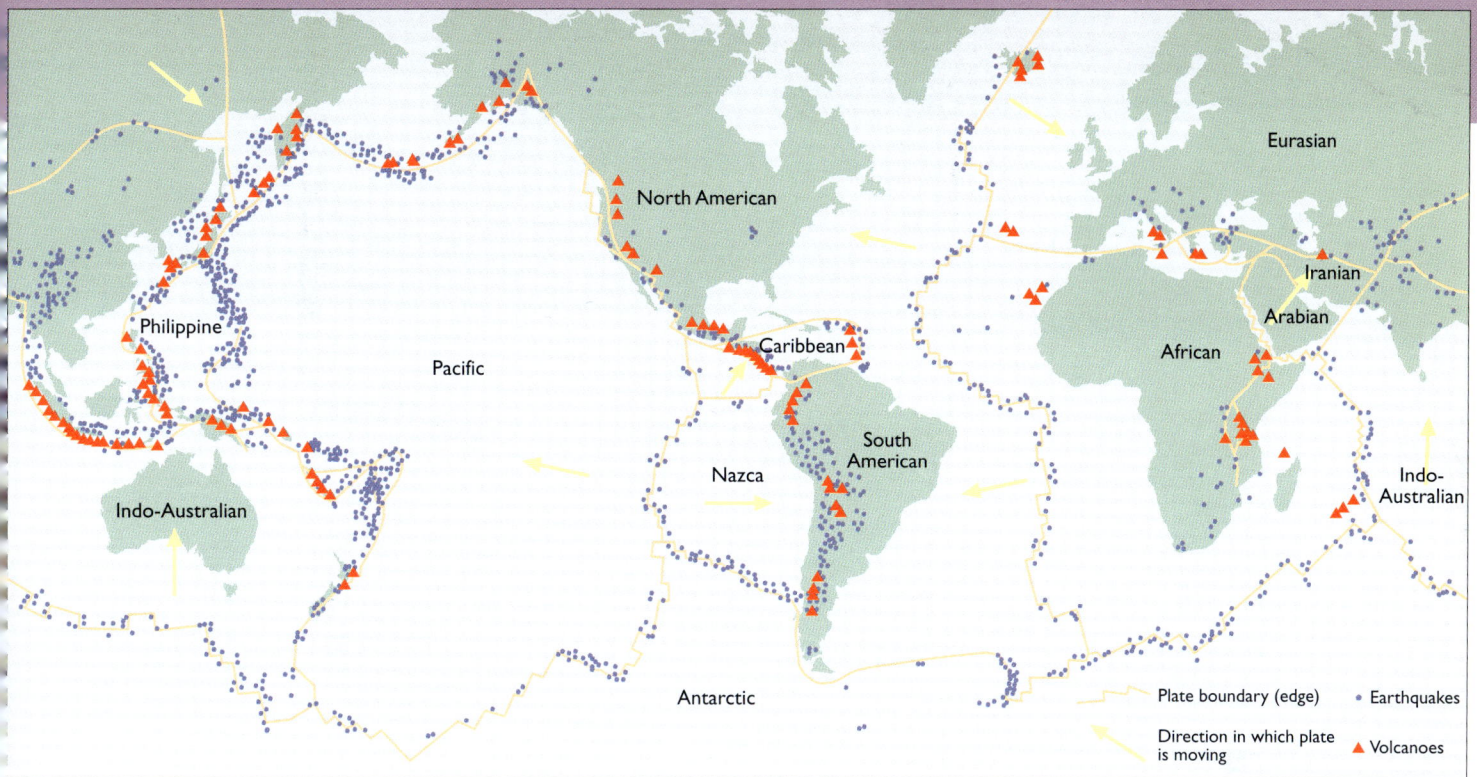

△ **Figure 3.8** Distribution of earthquakes, volcanoes and plate boundaries

The map (Figure 3.8) shows places in the world that are affected by either earthquake or volcanic activity. It is not a coincidence that the pattern made by the distribution of these events is almost exactly the same as the lines showing the boundaries between **tectonic plates**. As you have seen, there is a close link between tectonic plates and tectonic activity like earthquakes and volcanoes.

As we know where these events are likely to take place it should be an easy job to prepare for the problems they are likely to create – shouldn't it? Well in fact, even though we have a good idea of where the tectonic events might take place, it is still nearly impossible to know when things will happen.

ICT links

The following web addresses will help with your research into earthquakes and volcanoes.

www.pbs.org/wgbh/aso/tryit/tectonics/

http://volcanoes.usgs.gov/

http://earthquake.usgs.gov/

5W Activity

- **Hot spots**
- **Consructive plate boundaries**
- **Destructive plate boundaries**

WHAT are they?

WHERE do they occur?

WHY do they happen?

WHEN can they cause earthquakes and volcanoes?

WHO is most likely to see the results of these zones of activity?

Types of volcano

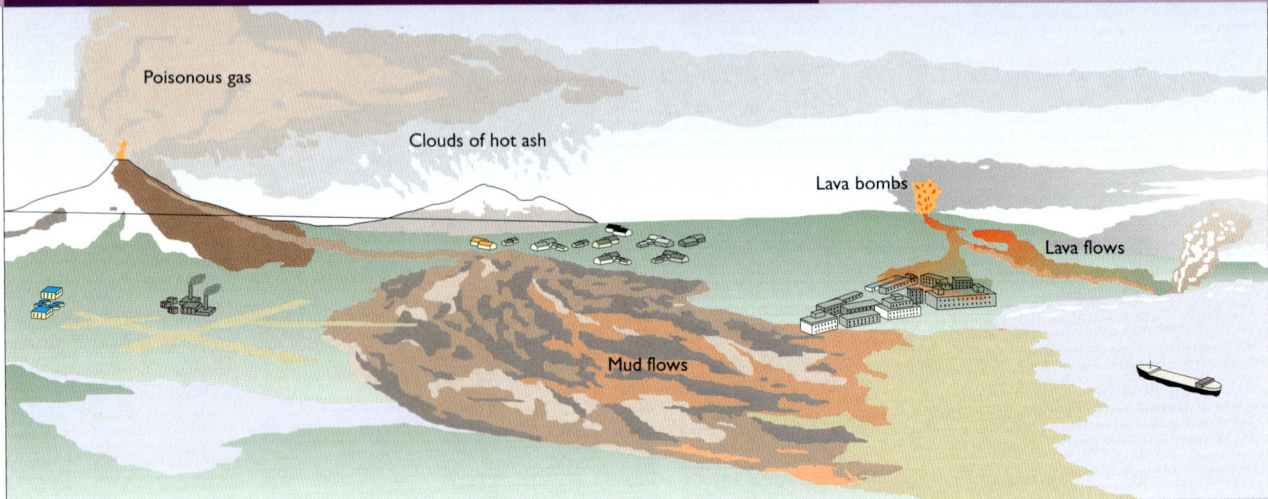

△ **Figure 3.9** Problems caused by volcanic eruptions

Volcanoes can produce a range of different material in their eruptions. As well as molten rock called **lava**, they can also erupt **poisonous gas**, clouds of very **hot ash** and **lava bombs**. The amount and type of material erupted depends on the type of volcano; they are not all the same!

Volcanoes which form at destructive plate margins usually form into **strato volcanoes**;

these produce the well known shape that people recognise as a volcano. At mid-ocean ridges and hot spots, it is more usual that a **shield volcano** will form. These are much flatter and cover a much wider area than Strato volcanoes. The lava that erupts from these volcanoes is often more liquid in its consistency, and flows quicker and further than other volcanoes, covering wide areas.

△ **Figure 3.10a**

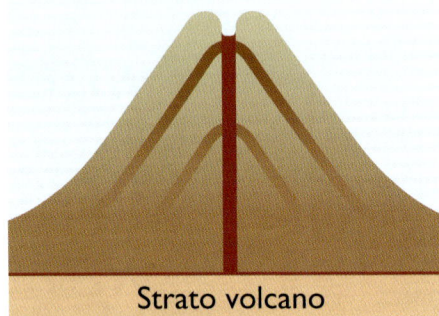

Strato volcano

△ **Figure 3.10b**

△ **Figure 3.11a**

Shield volcano

△ **Figure 3.11b**

Activities

1 Describe all the dangers that can come from volcanoes.

2 Which type of volcano is the most dangerous? Shield or strato? Explain your reasons.

About earthquakes

Earthquakes are vibrations caused by earth movements. An underground mine collapsing can cause a small earthquake. **Faults** in rocks slip suddenly. Most happen because of plate movements (Figure 3.12). Pressure builds up as the plates move and then suddenly a judder happens, triggering vibrations that can travel round the world.

The place in the crust where this earth movement happens is called the focus. The place on the surface above the focus is called the epicentre. Vibrations travel outwards from the epicentre like ripples on a pond (Figure 3.13). It is usually closest to the epicentre where the greatest damage can be done to people and the communities they live in. Away from the epicentre the impact of the earthquake decreases. With very sensitive equipment scientists can detect these shock waves many thousands of miles away.

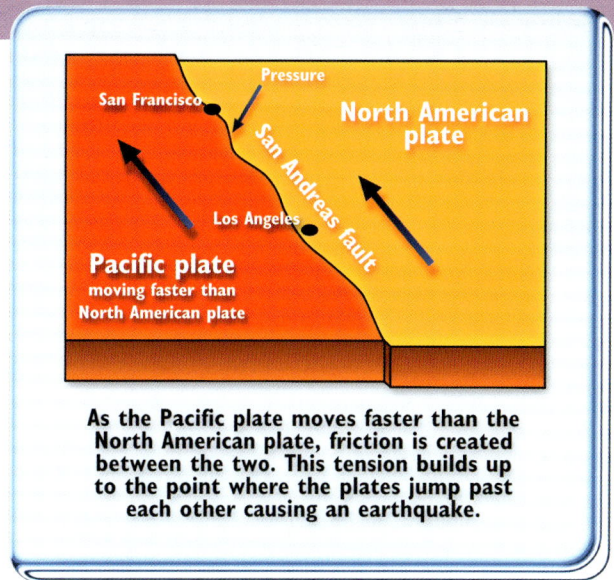

As the Pacific plate moves faster than the North American plate, friction is created between the two. This tension builds up to the point where the plates jump past each other causing an earthquake.

› Measuring the size of earthquakes

There are two different ways of measuring earthquakes. The **Richter Scale** measures the energy released by an earthquake at its focus. The **Mercalli Scale** describes the damage caused by an earthquake (Figure 3.14).

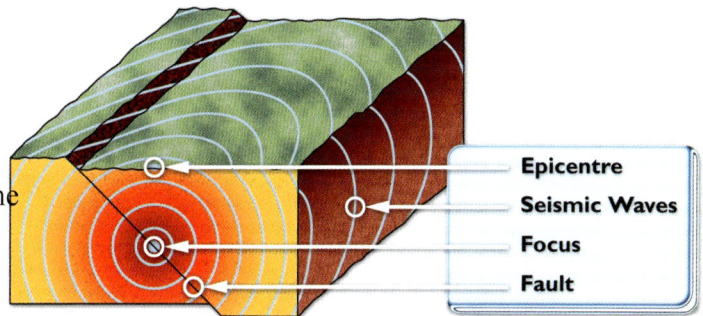

- Epicentre
- Seismic Waves
- Focus
- Fault

△ Figure 3.13

▽ **Figure 3.14**

Less than 3.5	3.5 – 5.4	Under 6.0	6.1 – 6.9	7.0 – 7.9	8 or greater
Generally not felt, but recorded	Often felt but rarely causes damage	At most slight damage to well designed buildings. Can cause major damage to poorly constructed buildings over small regions	Can be destructive in areas up to about 100 kilometres across where people live	Major earthquake. Can cause serious damage over large areas	Great earthquake. Can cause serious damage in areas several hundred kilometres across

Activities

1 When earthquakes are caused by suddenly shifting plates, does this make them easy or difficult to predict?

2 Which scale of measurement is more reliable when comparing earthquakes from around the world? Explain your answer.

3 What is the point in measuring the size of an earthquake?

4 Where is the safest place to shelter from an earthquake?

What is the impact of tectonic activity?

Having looked at the causes and types of volcanoes and earthquakes, it is now time to see how these events affect the people and the countries where they occur.

We are going to compare the effects of volcanoes and earthquakes in two different countries to see if there are similarities and differences in what happens as a result of a tectonic event.

The examples are from Japan and Peru. As you complete the comparison you should ask yourself: **'Do volcanoes and earthquakes affect rich and poor countries in the same way?'**

ICT links

To help complete the Extension tasks on page 55, it would be useful for you to do some research of your own into tectonic activity in Japan and Peru.

You could use the following websites to get started, or if you have a CD-Rom encyclopaedia, try using that.

www.your-nation.com

www.census.gov

FACT FILE: PERU

Popuation 26.11 million people
GDP $11.20 billion
Area 1.29 million km^2
Birth rate 26.69 births/1 000 people
Life expectancy 69.97 years
Literacy 88.70%

Source: CIA World Factbook 1998

FACT FILE: JAPAN

Popuation 125.93 million people
GDP $3.08 trillion
Area 377 835 km^2
Birth rate 10.26 births/1 000 people
Life expectancy 80 years
Literacy 99%

Source: CIA World Factbook 1998

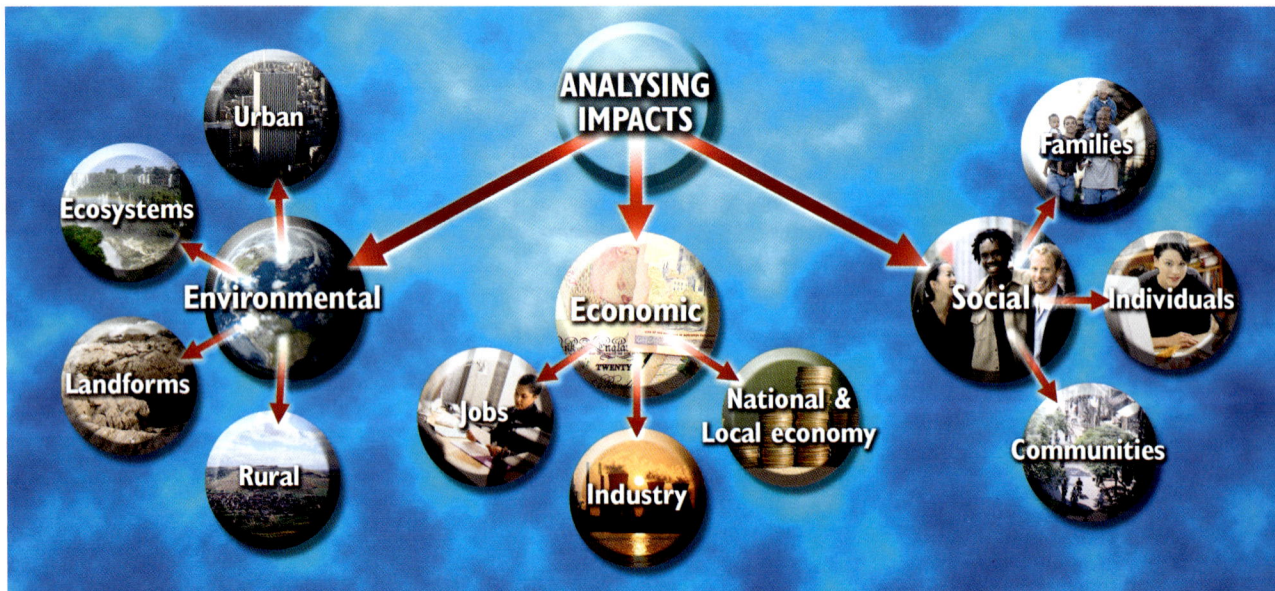

When comparing the impacts of events like earthquakes and volcanoes, it is useful to classify the effects using the headings in the diagram above.

△ **Figure 3.15**

Impacts of tectonic activity

Usu Erupts

Friday 31 March, 2000 – Japan's snow-capped Mount Usu volcano erupted today, spewing out a huge column of smoke and ash and hurling a flood of volcanic ash and rock toward a small nearby town.

Officials warned that the danger was far from over, however. 'It is possible that widespread damage could result from this eruption,' chief cabinet spokesman Mikio Aoki said from the government's emergency headquarters in Tokyo. Prime minister Keizo Obuchi was also at the headquarters monitoring the situation.

Roughly 51 000 people live in towns near Usu, 475 miles north of Tokyo on the island of Hokkaido. Experts had predicted the eruption because of increased **seismic activity** that began earlier this week, and the evacuation of more than 11 000 people began on Wednesday.

▷ **Figure 3.16**

△ **Figure 3.17**

Saturday 1 April, 2000 – Two more eruptions have rocked Mount Usu, the volcano in northern Japan, which started to erupt on Friday. The mountain is continuing to spew black smoke into the air, and scientists warn that increased seismic activity could mean another big eruption is imminent.

So far there have been no injuries, largely because people were evacuated well in advance. But some are fearful that they might never return to their homes, which are gradually being buried in volcanic ash and rock. At least 18 000 local residents have been evacuated from the area, and more than 3 000 soldiers and several naval vessels and helicopters are on standby to help with further evacuations. More than 15 000 people spent the night in temporary shelters set up in schools and public halls.

Saturday 8 April, 2000 – Parts of the town are destroyed everyday. Homes are covered in ash.

Television footage has shown moon-like landscapes covered in ash, roads twisted and cracked by underground pressure and seismic activity, as well as buildings and cars battered by falling rocks. Scientists said an eruption was likely to contain superhot lava, making it potentially more destructive than the previous ones.

The central government is sending 1 500 temporary housing units to the area to help shelter about 18 000 people camped out at evacuation centres, local officials said.

– adapted from the BBC News Website

Earthquakes

Thousands left homeless in Peru!

At least 97 people were killed on Saturday 23 June 2001 when a huge earthquake struck Arequipa.

The earthquake, measuring 8.1 on the Richter scale, was the worst to hit Peru for 30 years. At least 40 000 people in the south were affected, many of whom were made homeless. Peru's second biggest city, Arequipa, was the most severely affected. Most of the damage and casualties were around the city, where historic buildings were reduced to rubble. In the coastal town of Camaná, 39 people were killed by a tidal wave triggered by the quake. A powerful aftershock hampered the search for survivors.

From Ojo, Peru, 26 June 2001 – A great cloud of dust, walls destroyed, people walking about in the street, hungry and cold, this is the sad scene in this city. The devastating earthquake on Saturday left pain and desolation, villages almost erased from the map. The inhabitants of Moquegua overcoming their pain yesterday tried to retrieve any object or piece of furniture in any way useful. The town looked like it had been bombed, while despair spread among those affected for want of food and basic services.

From The Scotsman, 27 June 2001 – Forty-five minutes after Saturday's earthquake the first **tsunami**, 30 metres high, engulfed the mud huts of Camaná ... A second, bigger wave came moments afterwards leaving the once peaceful fishing town ripped to shreds ... Distressed villagers begged for help after their second night sleeping outdoors.

From Gestion, Peru, 26 June 2001 – Reconstruction in this part of the country will be much slower than in previous tragedies because of fewer financial resources. Political powers must respond generously to help victims of the south even if the economic recession of the last four years has practically paralysed industry, culture and commerce.

From La Republica, Peru, 26 June 2001 – It is clear that what is of interest now is to relieve the suffering of those affected, many of them living in new villages and in precarious buildings the majority of whom have lost everything. We must channel aid to them straight away.

Adapted from *The Guardian* 30th June 2001

It is not always easy to see or measure the impacts of these natural disasters, especially when we focus on the ways people have been affected. Although many of the people who live in Usu and Arequipa are aware of the risks of earthquakes and volcanoes, they still choose to stay. You would think that the problems caused by the earthquake and volcano in the two case studies would make people move away, but they haven't. Why is this?

The amount of knowledge and understanding that people have for the potential dangers of natural disasters is called Hazard Perception. The way people respond to an earthquake, volcano, or any natural hazard depends on their perception of the threat it poses to them. If they think their lives will be continuously put at risk, it is likely they will move.

This reliance on personal perceptions can be a problem in certain cases. People may think the risks of being affected by a natural hazard can be reduced by better emergency services, stronger building, or evacuation plans. In reality, while these things may help, there is nothing that can prevent natural disasters like volcanoes and earthquakes.

Less and More Economically Developed Countries (LEDCs and MEDCs) – Reminder!

Peru and Japan have different levels of economic development. Their income from exports helps with their ability to cope with the impacts of earthquakes and volcanoes.

So along with other things, a country's wealth is an important factor in dealing with these natural disasters.

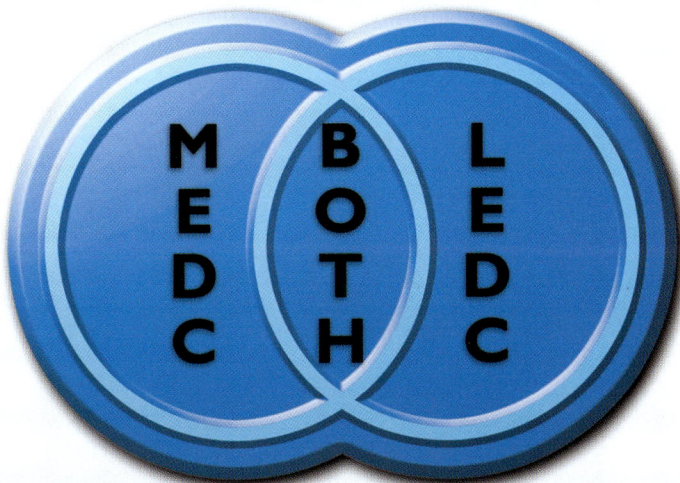

△ **Figure 3.18** Comparing the impacts of national disasters

Activities

Foundation

Comparing impacts

1 List the impacts of tectonic activity on both Japan and Peru.

2 Put your lists in order, with what you think are the greatest impacts at the top.

3 Explain why you have chosen the most important impact.

Target

4 Use the headings Social, Economic and Environmental to classify all the impacts listed in the two reports.

5 On a larger copy of the Venn diagram (Figure 3.18) sort out the impacts which are only relevant to Japan (MEDC), those only relevant to Peru (LEDC), and those which affect both.

Extension

6 Can you see a pattern in the type of impacts and the type of country affected?

7 If countries are affected by the same impacts, why does the number of victims often vary between LEDCs & MEDCs?

8 What part does wealth play in a country's ability to cope with natural disasters?

What is disaster relief and how can it help?

For many countries, both rich and poor, the results of a natural disaster like an earthquake can cause serious problems for months and even years after they happen.

Helping to make life easier for people affected by disasters are the international relief organisations. These are often charities which are funded by donations from individuals and companies. Providing the right kind of help at the right time can be a difficult job, so it is useful to sort the types of aid into categories. The easiest way to do this is shown in the diagram (Figure 3.20). People affected by earthquakes will have different needs at different times. Priorities change from the time the disaster happens, to weeks and months after. Immediate, emergency relief is the responsibility of the individual country, or even region. For those caught in the disaster, their most important needs are medical care, food, clean water and shelter.

With improvements in global communications, international aid can be available very quickly to the most remote locations, but what do they provide?

Along with emergency supplies, aid agencies can provide advice on the best ways to search for any people who are still missing and how to begin clean up operations. For even longer term help, they can help educate people in basic first aid techniques, or even give simple advice about creating stronger buildings. However, it is up to the country to plan to avoid future problems.

△ **Figure 3.19** People made homeless by an earthquake

△ **Figure 3.20** Different types of relief

▷ **Figure 3.21** Men inspecting rubble

A case study for disaster relief

Entire villages were leveled by the 7.9-magnitude tremor, centered in Gujarat state on 26 January, 2001. In the aftermath, hundreds of thousands of people were forced to sleep outdoors with little or no shelter, poor sanitary conditions, lack of food and inadequate clean water. Today, the huge task of housing reconstruction and rehabilitation has only just begun – leaving quake-traumatised survivors with the massive challenge of rebuilding their homes and shattered lives.

The Christian charity, **World Relief**, quickly responded to the immediate crisis, providing funds for distribution of blankets, tents, food, water and cooking utensils to outlying villages. In subsequent weeks, the emphasis shifted from providing immediate aid to long-term recovery and rebuilding – a process that could take years. As development work continues, World Relief plans to provide small business loans, called LifeLoans, to enable needy families to earn a living.

▷ **Figure 3.22**

△ **Figure 3.23** Wrecked building

Activities

1 Using the headings: Emergency Relief, Short-term Aid, and Long-term Aid, make lists of all the examples of aid described on these pages.

2 Which type of aid do you think LEDCs find it most difficult to provide themselves? Why?

3 Find out the name of a world aid agency and produce a multimedia presentation about the way it has helped countries around the world.

Disaster Relief Organisations

Disasters relief organisations are targeting over half a million people affected by the earthquake in Gujurat. Their aim is to care for health needs and prevent the spread of disease. The work has only just begun to help communities get back on their feet.

A spokesperson for Oxfam's Disasters Emergencies Committee said:

'There is so much for us to do! We have started thanks to the generosity of those people who have already given to the appeal. The people of Gujarat are trying hard to come to terms with this disaster which has destroyed their homes and livelihoods ...'

△ **Figure 3.24**

How can we prevent volcanoes and earthquakes from causing so many problems?

We will use the country of Japan to investigate what attempts people have made to lessen the impacts of both volcanoes and earthquakes.

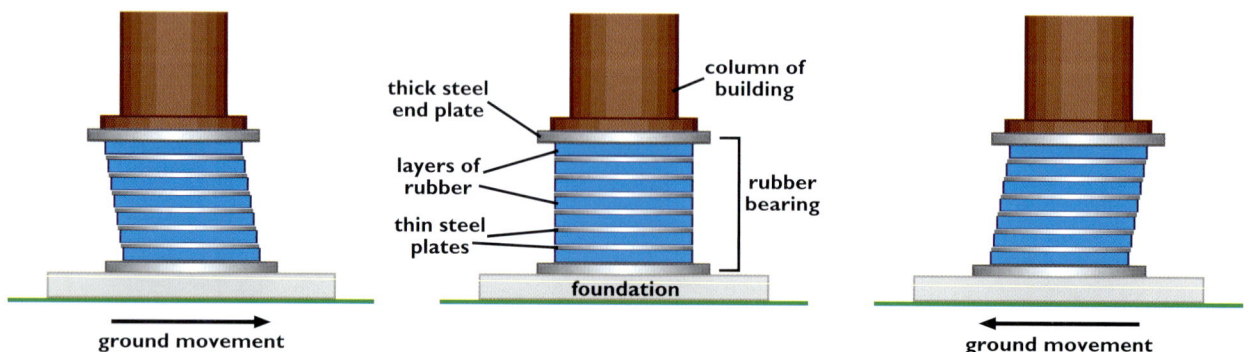

△ **Figure 3.25**

Japan has been affected by tectonic activity for many thousands of years. As you can see from the map (Figure 3.25), its position on the boundary between the Pacific and Eurasian plates means that earthquakes and volcanoes are a constant threat to the people of this nation. So what have they done to prevent these natural events turning into natural disasters, and has it worked?

The Japanese government has approached the problem in a way that can be summarised by this diagram.

▷ **Figure 3.26**

It tries to influence every stage of possible disasters, before, during and after the event. It has spent a huge amount of money on scientific research to try and predict when and where volcanoes and earthquakes are likely to happen. It hopes that if people get enough warning of these events, they will be able to get out of the danger zones.

Secondly, it teaches everybody in the country about the problems caused by volcanoes and earthquakes. This education starts at an early age, with every child being shown what to do in the event of a disaster. The people of Japan even have a day's holiday to practise their emergency drills.

Thirdly, many new buildings in the country have been designed so they can withstand certain amounts of movement from earthquakes. This is very expensive though and happens most in city centres where buildings have to be tall.

Finally, the country has spent a lot of money training and equipping its emergency services. Police, Fire and Ambulance teams are all prepared to work in the special conditions found after an earthquake or volcano.

▽ **Figure 3.27** Foundations designed to move in an earthquake

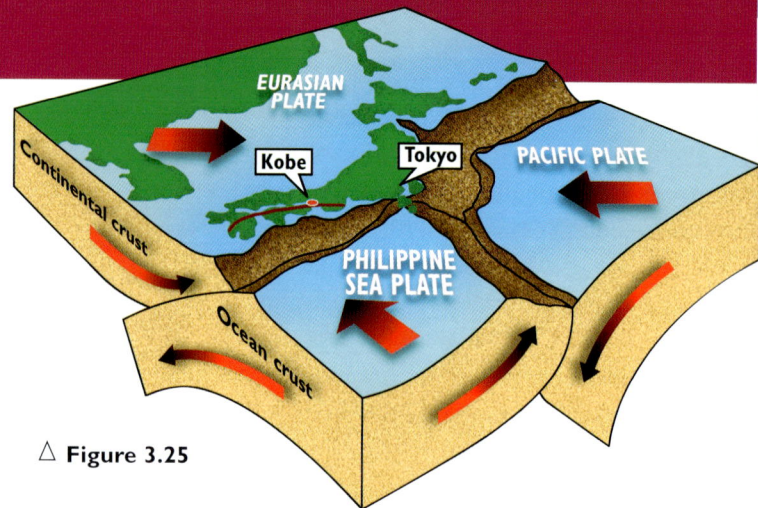

Eyewitness Report

The earthquake that hit Kobe on 17 January 1995 caused severe damage to the city and the surrounding areas. Hundreds of buildings collapsed, roads cracked, bridges were destroyed, there were several landslips, and water, power and telephone lines were severed.

Even though the emergency services had been well trained for a disaster like this, they found it very difficult to get to the worst hit areas. Without reliable water supplies it was difficult to keep fires under control.

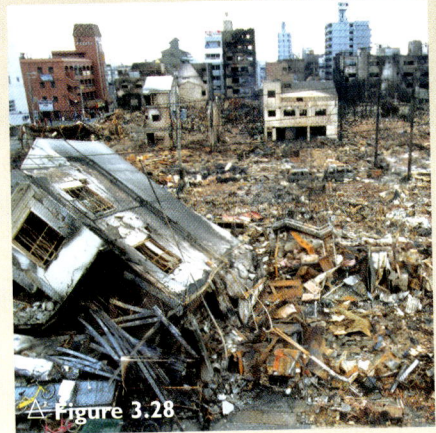
△ Figure 3.28

I had thought that all the news reports had prepared me for the extent of the damage but only when I was standing in the midst of 10 000m^2 of ashes and twisted metal did the scale of the disaster really hit me.

So many people died and the cost of rebuilding is so large, an estimated £200 billion. Only the most recent buildings in the city had been designed to withstand earthquakes and even some of these 'earthquake proof' buildings collapsed dramatically. We still have a long way to go before we can be really confident in our ability to overcome nature.

No amount of education and awareness could prepare people for such a devastating event. Even if they had predicted it people would have been caught trying to flee the city and deaths could have been higher.

△ Figure 3.29

Activities

Foundation

1 List three things that Japan does to prevent problems caused by earthquakes and volcanoes.

2 What were the effects of the Kobe earthquake on the city and its people?

Target

3 Read the '**eyewitness report**' of the Kobe earthquake and explain if you think the people of Japan were prepared for this disaster.

4 Is there anything that could have been done to reduce the number of people affected?

Extension

5 Is there any point in preparing for these natural disasters if we still cannot predict exactly when or where they will happen?

Philippine volcano spews truck-sized boulders

Tens of thousands of villagers have fled their homes as the Philippines' Mayon volcano unleashes a series of thunderous eruptions.

The volcano was spitting out flaming ash and huge great boulders. Witnesses said deafening booms rang out and giant cauliflower-shaped clouds of dust, ash and smoke shot up to 10 km into the sky, darkening this provincial capital of 120 000 people as well as surrounding towns.

The blasts prompted the Philippine Institute of **Vulcanology** and Seismology (PHIVOLCS) to raise the alert level around the volcano to a maximum five, meaning 'a hazardous eruption is in progress'. Officials said about 23 000 villagers fled their homes as the series of explosions, which began on Saturday night, intensified on Sunday and shook villages as far as 12 km away. 'The rocks coming down are as big as trucks,' vulcanologist Alex Baloloy said just before the first big blast at noon.

Rivers of fire – 'I heard what seemed like a huge thunder and I saw dark clouds ... then more boom-boom sounds,' local journalist Rhaydz Barcia said. Moments later, fiery rocks and gas thundered down the volcano's slopes at speeds estimated at 100 kph. PHIVOLCS chief Raymundo Punongbayan said the 'rivers of fire' bore temperatures of 900° Celcius, hot enough to incinerate anything in their path. Army trucks and police cars sped out of Legazpi

△ Figure 3.30

to threatened villages in a massive evacuation effort. Some villagers fled on carts driven by water buffaloes. In the panic, one woman collapsed with a heart attack while a pregnant mother prematurely gave birth, rescue officials said. One villager fleeing on a bicycle was killed when a speeding van carrying rescue teams struck him, officials said. Another truck full of evacuees fell into a canal, injuring some of them.

Overcrowding – They complain of overcrowding and a lack of facilities in the evacuation centres set up in nearby towns. Up to 15 families have been forced to share a room in converted schools, evacuees said.

△ Figure 3.31

There are also reports of shortages of food and water at the centres. Emergency workers, police and troops have been trying to stop people from heading homeward, but one relief worker said: 'No matter how much you try to stop them, they still wanted to go back.'

The precautions taken by the authorities ensured that nobody was killed as a direct result of the eruption, in spite of its violence – a remarkable achievement by local standards. The authorities are so confident of their ability to manage the effects of the eruption that they have even been trying to promote it as a tourist attraction.

Source: Erik de Castro, Reuters, Sunday 24 June, 01:26pm

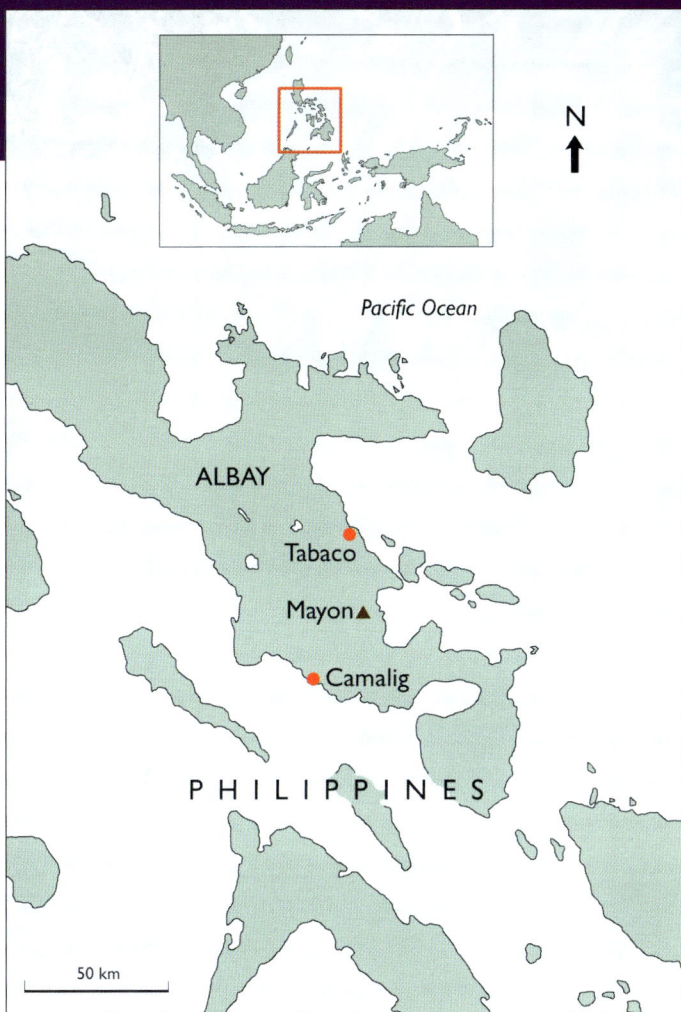

△ **Figure 3.32** Location map

▽ **Figure 3.33** Map of 1999 Mayon permanent danger and high susceptibility areas

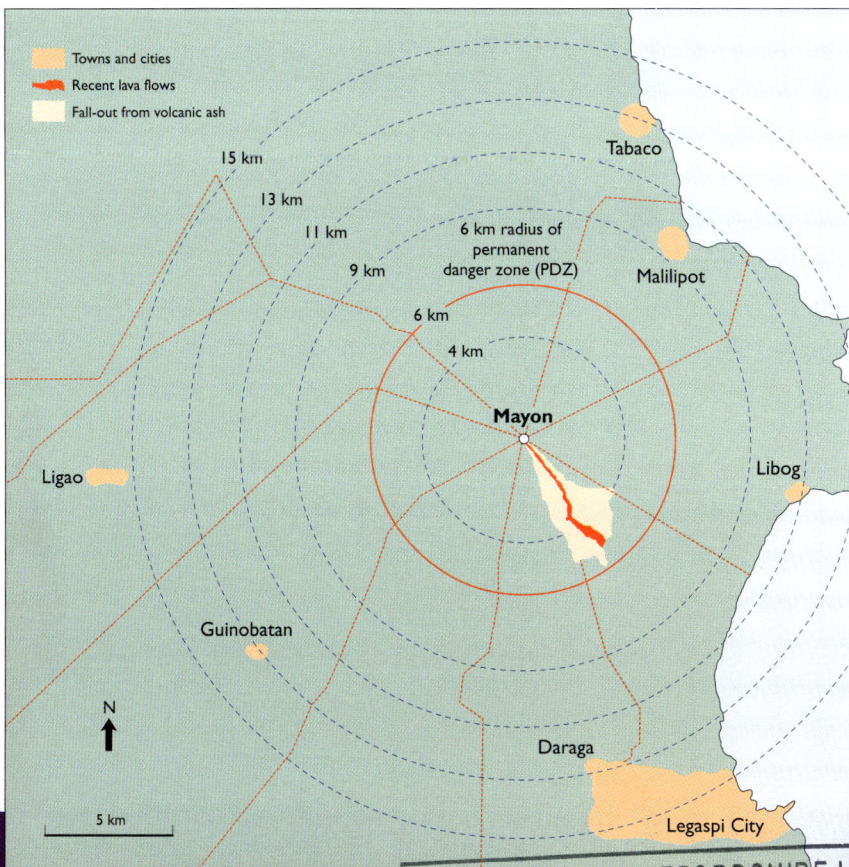

```
MAYON VOLCANO BULLETIN
(Released: 16 July 2001;
0800H)

ALERT LEVEL STATUS: 3
Remarks:
Alert Level 3 is still
in effect. However,
PHIVOLCS reminds the
public to avoid
occupying, on a long
term basis, areas within
the six kilometres'
radius Permanent Danger
Zone (PDZ) around the
volcano. Residents
living along the banks
of major channels are
advised to take the
necessary precautions
against lahars and
torrential stream flows
during unusually heavy
rains.
```

△ **Figure 3.34**

ICT link

www.phivolcs.dost.gov.ph/
Volcanoes/Mayon/MayonIndex.
html

Assessment tasks

Target task

Write a report to help the Philippines' government deal with the problems caused by the eruption of the Mayon volcano.

The report should contain an explanation for the eruption, backed up with relevant facts about the causes and effects. Make sure that you include diagrams, maps and pictures to illustrate your work. Include ideas for what could be done to improve the way eruptions are managed in the future.

For help, sort the questions in Figure 3.35 into a logical order before starting your report.

Who?
...has suffered the worst as a result of the eruption?
...can provide aid?

What?
...are the three most important things to do next?
...aid do you need immediately?

Why?
...did the eruption happen?
...is it so difficult to help everyone?

Where?
...is the danger area?
...can people stay if they are homeless?
...is aid needed most?

How?
...has the eruption caused other types of hazard, like mudslides?
...can you keep people away from the danger area?

When?
...can you move people back to their homes safely?

△ Figure 3.35

Extension task

Prepare a report and an action plan to help the Philippines' government deal with the problems caused by the eruption of the Mayon volcano.

The report should contain an explanation for the eruption, backed up with relevant facts about the causes and effects. Make sure that you include diagrams, maps and pictures to illustrate your work. The action plan should attempt to look at what has happened so far, what still needs to be done and how this experience can help improve our management of disasters in the future.

For help, sort the questions in Figure 3.35 into a logical order before starting your report and use a copy of Figure 3.36 for your action plan. Think carefully about the priorities for different groups of people, before, during and after the eruption.

Try to find out some additional information about this eruption by researching the websites listed in this chapter.

▽ Figure 3.36

Who?	Before	During	After
Scientists	Monitor the area, looking for any signs of increased volcanic activity		
Government			
Local People		Evacuate the area and stay in shelters set up by the government. Do not return until the area has been passed as safe again	

Review

This review page is designed to give you a chance to reflect on your understanding of this topic. The boxes at the top of each column represent your goal for each section of the topic. Start with the first column and read the statement in the bottom box. If you agree with it, move on to the box above. It is important that you are able to show your understanding at each level. An easy way to do this is to turn the statements into questions by switching the 'I can …' to read 'can I…?'.

If you want to move from one box to another but aren't sure how to, ask your teacher for help.

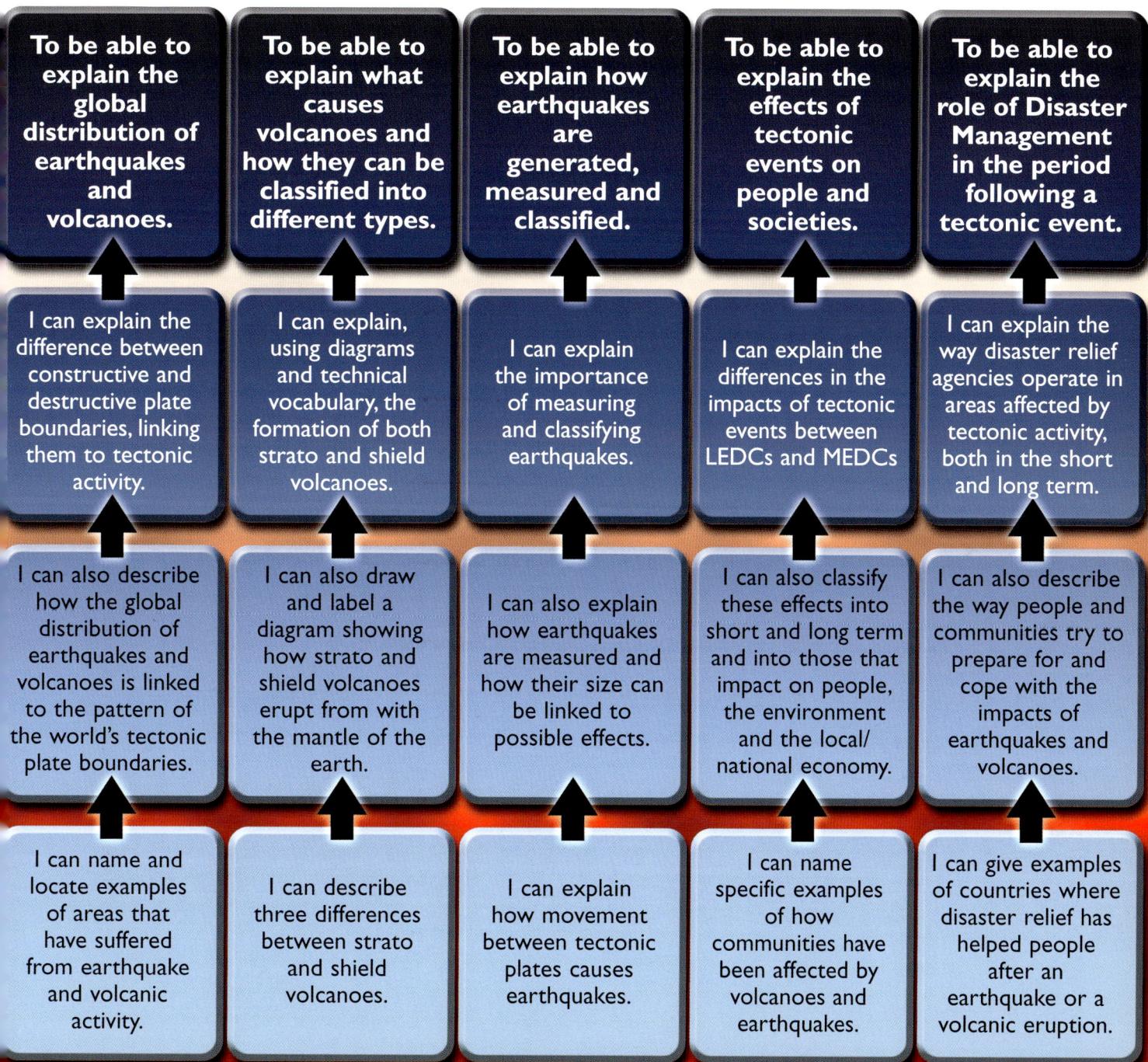

To be able to explain the global distribution of earthquakes and volcanoes.	To be able to explain what causes volcanoes and how they can be classified into different types.	To be able to explain how earthquakes are generated, measured and classified.	To be able to explain the effects of tectonic events on people and societies.	To be able to explain the role of Disaster Management in the period following a tectonic event.
I can explain the difference between constructive and destructive plate boundaries, linking them to tectonic activity.	I can explain, using diagrams and technical vocabulary, the formation of both strato and shield volcanoes.	I can explain the importance of measuring and classifying earthquakes.	I can explain the differences in the impacts of tectonic events between LEDCs and MEDCs	I can explain the way disaster relief agencies operate in areas affected by tectonic activity, both in the short and long term.
I can also describe how the global distribution of earthquakes and volcanoes is linked to the pattern of the world's tectonic plate boundaries.	I can also draw and label a diagram showing how strato and shield volcanoes erupt from with the mantle of the earth.	I can also explain how earthquakes are measured and how their size can be linked to possible effects.	I can also classify these effects into short and long term and into those that impact on people, the environment and the local/national economy.	I can also describe the way people and communities try to prepare for and cope with the impacts of earthquakes and volcanoes.
I can name and locate examples of areas that have suffered from earthquake and volcanic activity.	I can describe three differences between strato and shield volcanoes.	I can explain how movement between tectonic plates causes earthquakes.	I can name specific examples of how communities have been affected by volcanoes and earthquakes.	I can give examples of countries where disaster relief has helped people after an earthquake or a volcanic eruption.

Where should we decide to live?

People throughout the world will, at some time in their lives, have to make a decision about where to live. Here are some examples.

△ Figure 4.1

△ Figure 4.2

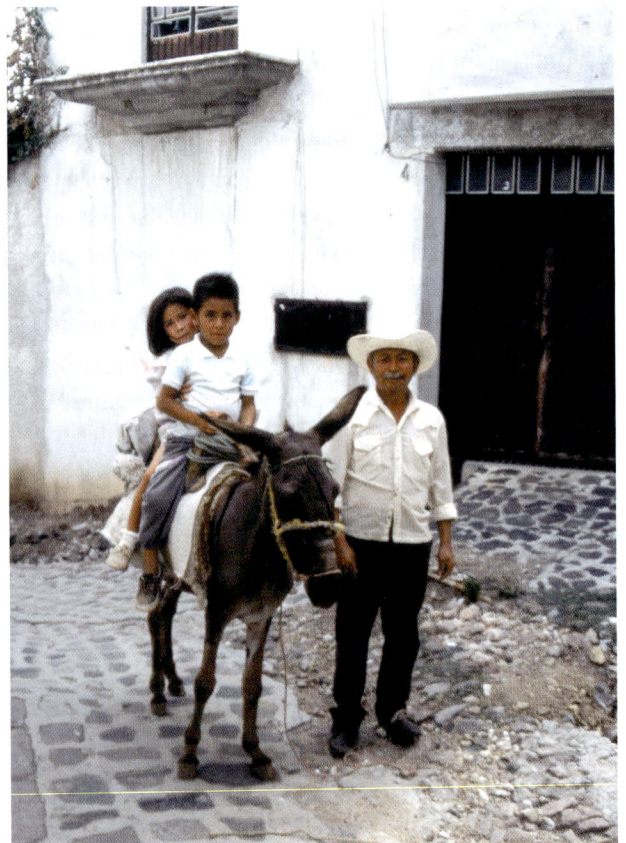

△ Figure 4.3

Activities

1 Look at Figures 4.1, 4.2 – 4.3 – but not the boxes below!

a What do you think these people have in common about their lives?

b What do you think is different about their lives?

2 Now read the fact boxes below.

Dan and Amanda moved into their village cottage in Hilderstone in the north Midlands. They moved from the nearby town of Stone. They were both born in large cities and **migrated** to the village four years ago. They bought up part of an old farmhouse and have restored it into a family home. They have two young sons who are driven to school in a nearby village. They like the fresh air and quiet of the countryside but both work in local towns. Many people in Britain are now moving out of towns to live in nearby villages. This is called **counter-urbanisation**.

These two men were born in a very poor village in West Africa. The photograph is over 100 years old. They are seamen. They worked hard for their living as stokers. This is a dirty and hot job shovelling coal into ships' boilers. They worked on British steamships working out of Cardiff. They settled here because it was easier to get a job and there were people from their own country living in Cardiff. They were **international migrants**.

Benito is the grandfather of José and Manuela. They have lived all their lives in the village of Tehuilotepec 5 km north of the historic small town of Taxco (population 95 000), which is situated 170 km south west of Mexico City. Their father has moved to Mexico City in the hope of a better job, and he plans to bring his family there when he has made enough money. He would be called a **rural–urban migrant**. Benito plans to live out the days of his life in the village where he was born.

a Who has moved from a large city to a village? Why did they move?

b Why do you think the African men did not stay in the village of their birth?

c What will the future of José and Manuela be? Where do you think they will live as adults? Why do you think their grandfather is not planning to move?

d Why do you think people in Britain are moving into country areas but people in Mexico are moving away from them?

Is home sweet home?

From earliest times humans have needed to shelter by building a **settlement**. It can vary in size from a single house to a large city or **conurbation**. In some parts of the world many people live in single houses not close to others. These areas are often very **rural**. Most people will live close to other houses, forming larger settlements. A small collection of rural houses is called a **hamlet**. A **village** is formed when more houses cluster and a few services are provided, such as a shop. Most people now live in **urban** areas in large towns or cities. Here there are more services and **amenities** than you would find in a village, such as a cinema. There are also better job opportunities.

People's experiences of settlements differ throughout the world

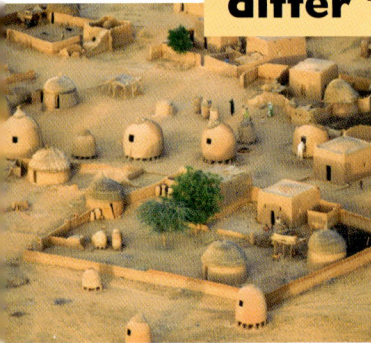

△ **Figure 4.4** Desert village, Niger 15° 03' N, 5° 12' E

△ **Figure 4.5** Borth village, central Wales, 52° 59' N, 4° 3' W

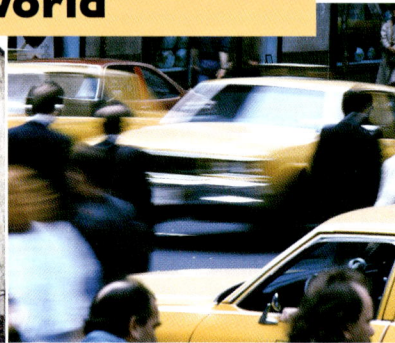

△ **Figure 4.6** New York street scene, 40° 45' N, 74° 0' W

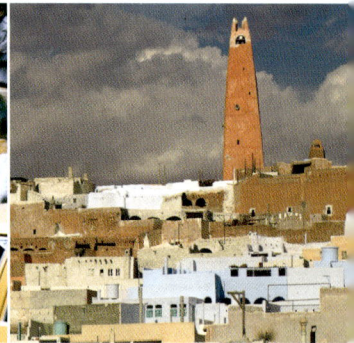

△ **Figure 4.7** African city, Ghardaia, 32° 31' N, 3° 37' E

Try this method of looking at photographs in a different way. For each first letter of the four main compass points you are encouraged to write about what you see.

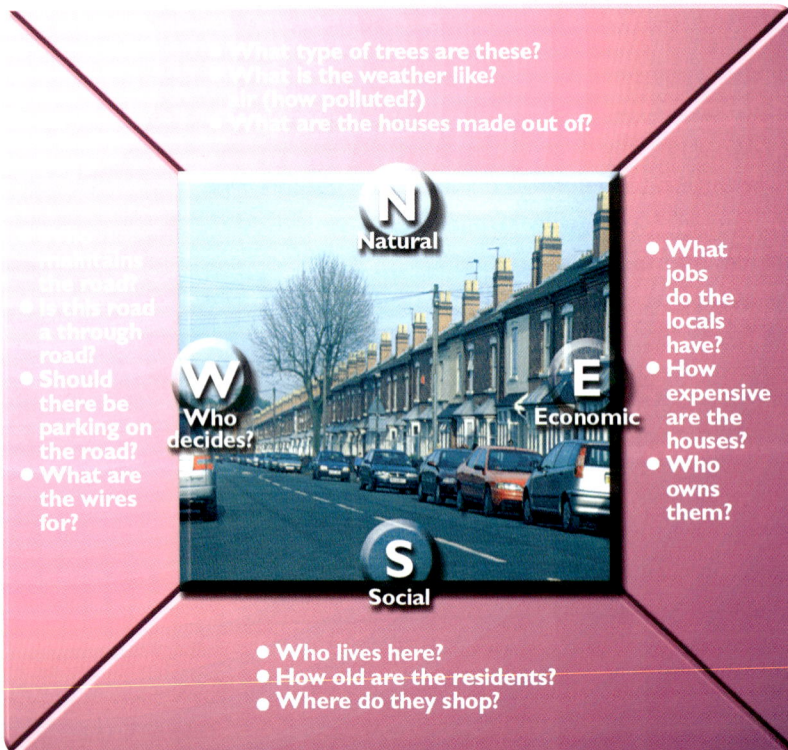

N Natural
- What type of trees are these?
- What is the weather like?
- Air (how polluted?)
- What are the houses made out of?

W Who decides?
- Is this road a through road?
- Should there be parking on the road?
- What are the wires for?

E Economic
- What jobs do the locals have?
- How expensive are the houses?
- Who owns them?

S Social
- Who lives here?
- How old are the residents?
- Where do they shop?

△ **Figure 4.8** Compass rose diagram

Activities

Figures 4.4–4.7 show four settlements in different parts of the world.

1 Which ones are rural and which are urban?

2 What types of services are available nearby?

3 Which ones are in **MEDCs** and which are in **LEDCs**?

4 Why do you think they are so different? Do they have any similarities?

5 What would you like and dislike about living in these settlements? Do you think the people living there will have the same ideas as you?

Will city life be for all?

Country	% population urban	growth rate % per year urban population	telephones per 1000 population	Workforce percentages		
				agriculture	industry	services
Australia	84.7	1.2	496	6	26	68
Burkina Faso	16.4	9.8	3	92	2	6
India	27.1	3	11	64	16	20
Japan	78.2	0.4	480	7	34	59
Mexico	73.6	2.5	93	28	24	48
Nigeria	40.5	5	3	43	7	50
UK	89.3	0.4	488	2	29	69
USA	76.3	1.2	602	3	26	61

△ **Figure 4.9** Selected statistics from countries at different levels of development

Activities

Foundation

1 In Figure 4.9, which country has the greatest percentage of people living in towns and cities?

2 Which country has the greatest percentage of people living in villages?

3 Which country's towns and cities are growing the fastest?

Target

4 Can you see a relationship between the percentage of urban population and the percentage employed in agriculture? Devise a way of displaying this relationship as a graph.

Extension

5 As a country urbanises, other changes take place. Use the statistics about telephone ownership and the workforce (Figure 4.9) to describe what some of these changes are.

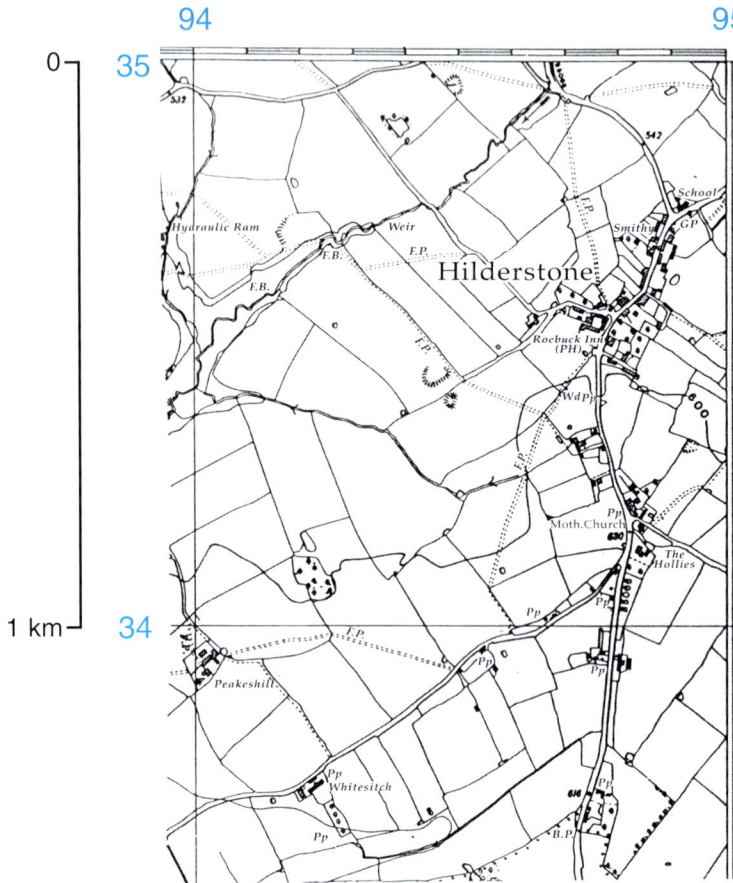

△ **Figure 4.10** Hilderstone 1949

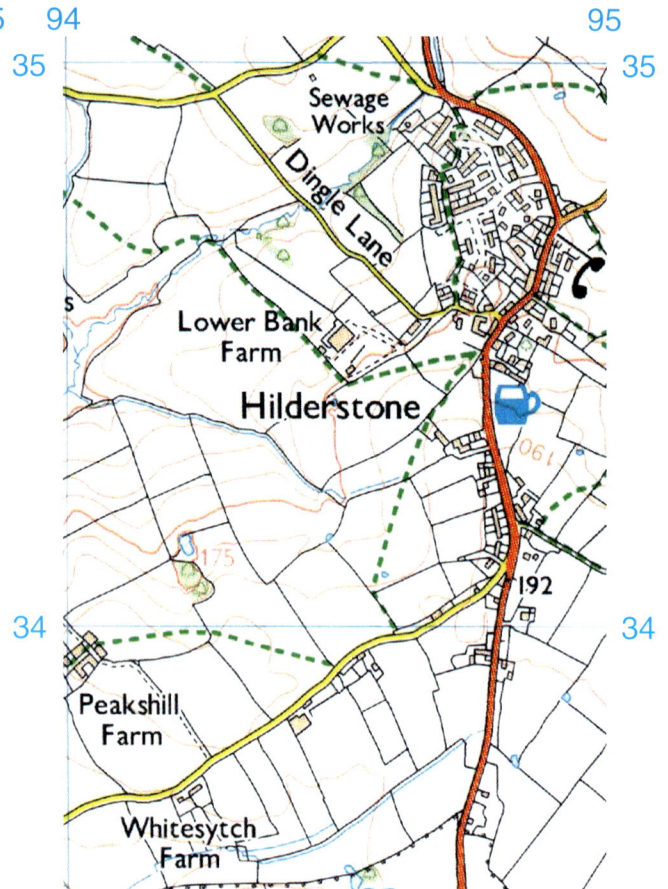

△ **Figure 4.11** Hilderstone today, from OS   © Crown copyright 2002

Hilderstone is a village with a population of over 400. It is situated in the north Midlands of England, 11 km to the south east of the city of Stoke-on-Trent. There are a number of **pastoral** farms in the area but like many other villages the majority of people do not have any connections with farming. Many now commute to nearby towns. Things were different years ago when most people either farmed or served the needs of the local community.

COMMERCIAL.

Barker Frederick, shoe maker
Bloor John, farmer, Neville cottage
Bossen John farmer, Peak's hill
Bowers William, farmer
Bridgwood William, farmer
Cheadle Henry, farmer
Cope William, Horseshoe P.H. & carrier

Dalton George, tailor
Dalton Joseph, tailor
Fairbanks Alice (Mrs), shopkeeper, Sharpley heath
Fairbanks Wm. brick ma. Sharpley hth
Harthan Samuel, farmer, Stone heath
Heath Tomas, farmer, Stone heath
Heath William, sawyer, Stone heath
James George, wheelwright
Johnson William, farmer
Leedham Arthur William, farmer & landowner, The Lessows
Mayer George, farmer, Newfields & at Day hills
Meddings George, grocer, Post office
Meddins George, farmer, Green farm
Mountford John, farmer

Phillips Thomas, blacksmith
Porter Thomas, ownkeeper
Price William, farm baliff to John Bourne esq. jun. J.P.
Sharratt William, farmer, Stone heath
Shelley Thomas Holdcroft, Bird-in-hand P.H. Sharpley heath
Till George, bricklayer
Tunnicliff Lydia(Mrs.),butcher & farmer
Tunnicliff Thos. B. farmer, Whiessich
Turner Thomas, Roebuck inn
Udall George, farmer
Urion Daniel, farmer, Spot grange
Urion Frederick, farmer, Wooliscroft
Vernon John, farmer, Spot farm
Walkerdine Samson, farmer

Figure 4.12a shows part of a **directory** for the village for 1892. They are useful for finding out about what people did and what services were available. A hundred years later the 1991 **Census** (Figure 4.12b) shows a different picture. Farming is still important for the village and surrounding area but there is a different mix of jobs today.

△ **Figure 4.12a** Directory for village of Hilderstone, 1892

▽ **Figure 4.12b** Employment in Hilderstone, 1991

Employment	Agriculture	Energy & Water	Mining	Construction	Banking & Finance	Government & other services
Percentage	21%	7%	7%	27%	7%	31%

◁ **Figure 4.13** Church

△ **Figure 4.14** Barn conversion

▽ **Figure 4.15** Bored children

1 Study Figure 4.12a. Make a tally chart listing the jobs in Hilderstone 1892.

2 Compare the type of jobs found in the village a hundred years ago with today.

3 Some of the 1892 job types will not be found in Hilderstone today. Why have they disappeared?

4 Look at the two maps for Hilderstone 1949 and today (Figures 4.10 and 4.11). Make a list of five changes that you notice between the two maps.

5 What three services are shown by **map symbols** on the more recent map?

6 Can you find any evidence of changes to farming in this area?

7 Look at the photograph (Figure 4.14) that shows an old farm's barn in the village converted into a house. What will be the good and bad effects of this action for the village?

8 Young people in the village think that life here is boring! Are there any activities that can be carried out here by teenagers that cannot take place in towns and cities?

9 The Parish Council would like to make a parish trail around the village. Starting at the crossroads (grid reference 949347) devise a trail using directions, distances, and map references for a walk around the village. Point out any changes that have taken place in the last 50 years.

▽ **Figure 4.16** Bus timetables

8A
First PMT

HANLEY – LONGTON – BLYTHE BRIDGE – HILDERSTONE

HANLEY Old Hall Street (return via Lichfield Street, Birch Terrace to HANLEY Bus Station), Lichfield Street, Victoria Road, Fenton, King Street, LONGTON Market Street (return via The Strand), Uttoxeter Road, Meir, Uttoxeter Road, Catchems Corner, Lysander Road, Farnborough Drive, Grindley Lane, Blythe Bridge, Draycott, Cresswell, The Hunter, Saverley Green, Fulford, Townend, Cross Gate, Moss Lane, Mossgate, Bird-in-Hand, HILDERSTONE.

Monday to Saturday

		NS	NS	S	NS		
		0841	1041	1241	1341	1541	1741
HANLEY, Argos Superstore	....	0856	1056	1256	1356	1556	1756
LONGTON, Market Street (Stop F)	....	0902	1102	1302	1402	1602	1802
MEIR, Broadway	....	0907	1107	1307	1407	1607	1807
BLYTHE BRIDGE, Rail Station ≈	0732	0910	1110	1310	1410	1610	1810
Draycott, Church	0735	0913	1113	1313	1413	1613	1812
Cresswell, Rookery Crescent	0738	0915	1115	1315	1415	1615	1815
Saverley Green, Greyhound	0740	0918	1118	1318	1418	1618	1818
Fulford, Village Hall	0743	0920	1120	1320	1420	1620	1820
Mossgate, Moss Lane	0745	0925	1125	1325	1425	1625A	1825
HILDERSTONE, Old School	0750						

	S	NS		S	1034B	NS		S	NS	
HILDERSTONE, Old School	0749	0754	0934	1034B	1134	1334	1434	1634B
Mossgate, Moss Lane	0753	0758	0938	1038	1138	1338	1438	1638
Fulford, Village Hall	0756	0801	0941	1041	1141	1341	1441	1641
Saverley Green, Greyhound	0759	0804	0944	1044	1144	1344	1444	1644
Cresswell, Rookery Crescent	0801	0806	0946	1046	1146	1346	1446	1646
Draycott, Church	0803	0808	0948	1048	1148	1348	1448	1648
BLYTHE BRIDGE, Rail Station ≈	0807	0812	0952	1052	1152	1352	1452	1652
MEIR, Broadway	0813	0818	0958	1058	1158	1358	1458	1658
LONGTON, The Strand (Stop B)	0820	0825	1005	1105	1205	1405	1505	1705
HANLEY, Bus Station	0835	0845	1025	1130	1225	1425	1525	1725

Notes:
A – Saturday only continues to Milwich
B – Saturday only commences from Milwich at 1630
NS – Not Saturday
S – Saturday only

Most journeys extend to/from Chell via High Lane

Sunday and Bank Holidays: No service

Page 12

249
TAXICO

HOLLINGTON – FREEHAY – CHEADLE – BLYTHE BRIDGE – HILDERSTONE – STONE

HOLLINGTON, Winnothdale, FREEHAY, Rakeway, Rakeway Road, Park Avenue, Tean Road, Tape Street, CHEADLE High Street, The Terrace, Town End, The Green, A521, Forsbrook, Cheadle Road, BLYTHE BRIDGE Uttoxeter Road, Draycott, Cresswell Lane, The Hunter, Saverley Green, Saverley Green Road, FULFORD, Fulford Road, Mossgate, MossLane, Bird-in-Hand, HILDERSTONE, B5066, B5027 Uttoxeter Road, Dayhills, Milwich, Dayhills, B5027 Uttoxeter Road, Little Stoke, Lichfield Road, Lichfield Street, Stafford Street, STONE Crown Street, Radford Street, Christchurch Way (returns direct from Christchurch Way to Lichfield Street).

Certain journeys operate via Tean Road, Tape Street, Chapel Street, Bank Street, Watt Place, CHEADLE High Street, Tape Street, Tean Road, A522, Upper Tean, Draycott Road, Draycott, Cresswell Lane, The Hunter, Bird-in-Hand, HILDERSTONE, B5066 and B5027 Uttoxeter Road between Park Avenue and Little Stoke.

Tuesday and Friday only

		F	
HOLLINGTON, Village Hall	0930F	1150
Winnothdale	0935F	1155
Freehay, Village Hall	0940F	1200
CHEADLE, High Street	0948	1208
Upper Tean, High Street		1215
Forsbrook, Square	0958	
BLYTHE BRIDGE, Duke of Wellington	1000	
Draycott, Church	1003	1218
Cresswell, Rookery Crescent	1004	1219
Saverley Green, The Greyhound	1006	
Fulford, Village Hall	1009	
Mossgate, Moss Lane	1011	
Hilderstone, Old School	1015	1226
Milwich, The Allways	1020	
STONE, Christchurch Way	1030	1235

			F	
STONE, Christchurch Way			1105	1235
Milwich, The Allways			1114	1245
Hilderstone, Old School				1250
Mossgate, Moss Lane				1254
Fulford, Village Hall				1256
Saverley Green, The Greyhound				1259
Cresswell, Rookery Crescent			1121	1301
Draycott, Church			1122	1302
BLYTHE BRIDGE, Duke of Wellington				1305
Forsbrook, Square				1307
Upper Tean, High Street			1125	
CHEADLE, High Street			1132	1317
Freehay, Village Hall			1140	
Winnothdale			1145	
HOLLINGTON, Village Hall			1150	

Notes:
F – Friday only
▢ – Service operated under contract to Staffordshire County Council

For alternative journeys between Blythe Bridge and Hilderstone, see Service 8A timetable

On other days of the week: No service

Page 22

▷ **Figure 4.17** Bus map

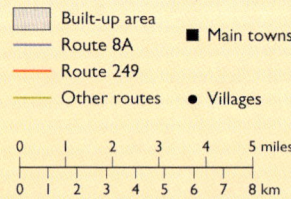

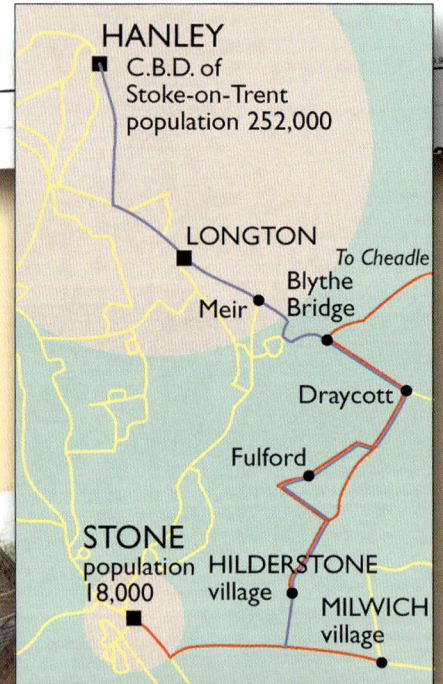

N

Key:
- ▨ Built-up area
- — Route 8A
- — Route 249
- — Other routes
- ■ Main towns
- ● Villages

| 0 | 1 | 2 | 3 | 4 | 5 miles |
| 0 | 1 2 3 | 4 5 | 6 7 | 8 km |

HANLEY
C.B.D. of Stoke-on-Trent population 252,000

LONGTON
To Cheadle
Blythe Bridge
Meir

Draycott
Fulford

STONE
population 18,000

HILDERSTONE village
MILWICH village

△ **Figure 4.18** A long wait?

8A City Centre, Blythe Bridge
HILDERSTONE

First

M382 SRE

▷ **Figure 4.19** The village bus arrives at last!

The village of Hilderstone has two bus services to the local market town of Stone and the city of Stoke-on-Trent. Figure 4.16 shows the buses serving the village along with a timetable. The nearest town is Stone (population 18 000), which has two supermarkets, banks and a small range of clothes and shoe shops. The nearest city is Stoke-on-Trent (population 252 000) with its **Central Business District (CBD)** at Hanley. Here we find a large covered shopping centre with all the major shops found in our big cities. Hanley also has banks, cinemas and theatres. It is the main shopping area of North Staffordshire. There is a bus connection to the mainline rail station, which has hourly links to London.

Hilderstone has a population of about 400. The village does not have a shop or post office. 11% of households do not own a car but half have more than one car. Nearly three quarters travel to work by car. Some people **car pool**, which means sharing a car journey to work. 13% of the people in this village and the surrounding area work at home. Many **telework** using telephone, computers and email.

The Government thinks that villages should have a better bus service. They have given **subsidies** to provide more buses and more routes. People are coming up with new ideas to improve rural bus services, e.g. in Wiltshire, an English county with plenty of villages, the *Wigglebus* runs along a main route but 'wiggles' off route to pick up passengers. Since its launch, 2 500 more passengers use the bus.

▽ **Figure 4.20** Teleworker

Foundation

1 What time is the earliest bus you can catch to Hanley on a weekday?

2 What time is the last bus you can catch from Hanley on that day?

3 Which days can you not take a bus?

4 Who will be using the bus from this village?

Target

5 How long does the bus take from Hilderstone to Hanley and from Hilderstone to Stone?

6 How many kilometres from Hilderstone will you travel on the bus for these two journeys?

7 Why are the buses so small?

Extension

8 What is the average speed for a journey to
a Hanley, and
b Stone?
You will need to work out time and distance and use the equation

$$\frac{\text{Time}}{\text{Distance}}$$

9 How does this compare with car travel times?

10 What would you suggest to increase the use of buses from this village?

What changes are taking place in the city?

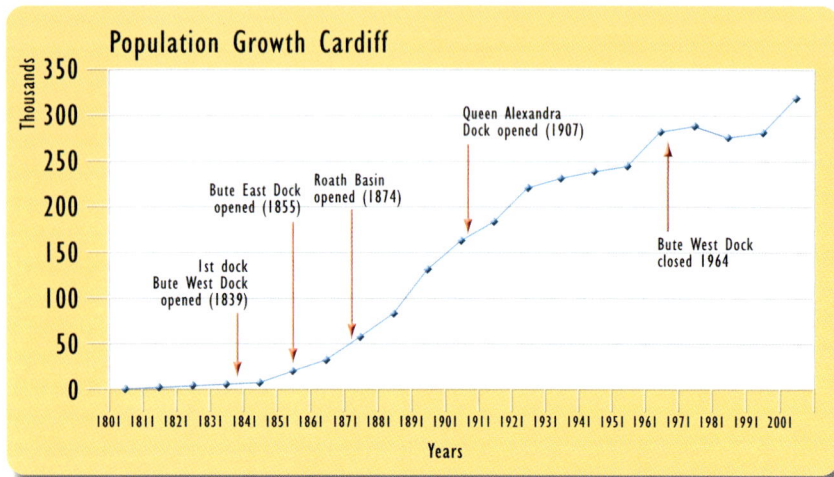

△ **Figure 4.21** Population growth graph, Cardiff

▷ **Figure 4.22** Tiger Bay, Cardiff

△ **Figure 4.23** Bute Street, Butetown, before re-development

▷ **Figure 4.24** Many local residents worship in the mosque. This is an old mosque, one of Britain's first. This photo is from the 1970s

City life has had big changes. Here is an example of how one area of a city has undergone rapid change in recent years. The city is Cardiff (population of over 300 000), situated in south Wales.

A city of immigrants

Cardiff rose to fame when docks were built for the export of coal. As a result, many people migrated from all parts of Britain in the 19th century to fill the new jobs that were created in this growing seaport. Many of the **immigrants** came from villages seeking a better life away from the poverty of the countryside. Cardiff also attracted African sailors who settled in an area close to the docks called Butetown. This area was also nicknamed 'Tiger Bay', by the many seamen who visited the port from all over the world.

Decline of the docks

Cardiff docks stopped exporting coal in the 1960s, a steelworks was closed in the late 1970s, and the **residential** area of Butetown became run-down. Old houses occupied mainly by descendants of African sailors were considered unfit, and the area of Butetown was re-developed. Two new high-rise blocks of flats were built as well as other high-density housing. A new park was formed with the filling in of the Glamorganshire canal. A new mosque was built to serve many of the members of the Somali community who lived in this area. Unemployment reached as high as 50% for males.

The Cardiff Bay story

In 1987 a move was made to try and re-develop the run-down dock area. The Cardiff Bay Development Corporation was set up with the task of regenerating the area. To achieve this the following was carried out:

- A 1.1 km barrage built across the Taff and Ely estuary to create a freshwater lake.
- A new road built to link the area with the M4 motorway.
- A new avenue built to link the area with the City Centre to the north.
- New industry attracted (30 000 new jobs planned) ranging from manufacturing television tubes to financial jobs in major banks.
- New housing built in the area, such as Atlantic Wharf in Bute East Dock.
- Siting the Welsh Assembly, the government for Wales.
- Leisure and tourism encouraged around the Inner Harbour with restaurants, a science museum called Techniquest and a health spa hotel.

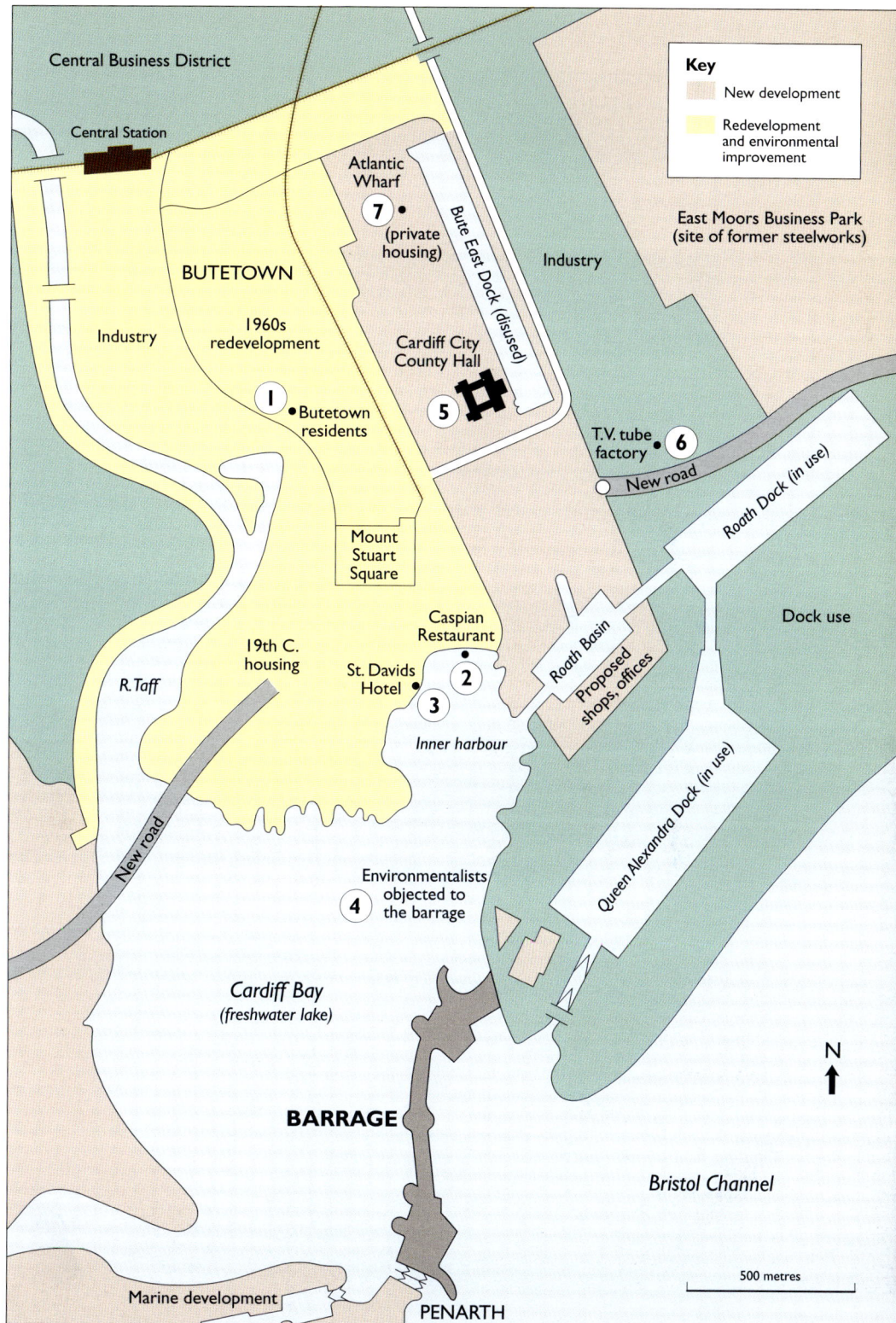

Key
- New development
- Redevelopment and environmental improvement

Central Business District

Central Station

Atlantic Wharf

7 (private housing)

BUTETOWN

1960s redevelopment

Industry

1 Butetown residents

Bute East Dock (disused)

Industry

East Moors Business Park (site of former steelworks)

Cardiff City County Hall

5

T.V. tube factory **6**

New road

Roath Dock (in use)

Mount Stuart Square

Caspian Restaurant

St. Davids Hotel

2

3

Roath Basin

Proposed shops, offices

Dock use

19th C. housing

R. Taff

Inner harbour

New road

Environmentalists objected to the barrage

4

Queen Alexandra Dock (in use)

Cardiff Bay (freshwater lake)

BARRAGE

N

Bristol Channel

500 metres

Marine development

PENARTH

△ **Figure 4.25** Map of Cardiff Bay

What changes are taking place in the city?

Has the re-development been a success?

Some people have raised doubts about the way the Cardiff Bay scheme has developed. Some locals have said, *'It's not for us ... we get few jobs ... we can't afford to go to the pricey restaurants'*. Others have felt that the development is the best thing to happen to the area *'before this we had mudflats, **derelict** buildings and a poor image ... it's put the area on the map. No more Tiger Bay but the new Cardiff Bay!'*

Activities

This can be carried out as a class activity, or individually.

1 Choose one of these seven groups interested in the Cardiff Bay area.

Manager of St. David's Hotel

This is Wales' first five star hotel with a health Spa. It's a prestige location.

Butetown residents

We already lived here before the scheme. Male unemployment of 50% is one of the highest in Wales.

Environmentalists

We never wanted the barrage in the first place. Many seabirds do not have the mudflats to feed from.

Owner of a new Caspian restaurant

I have rented a purpose-built building built over water with good views.

Atlantic Wharf residents

We moved here because we wanted to be near the city centre. The houses and flats are new but we will probably move on in a few years time. We do not know any local people as they live the other side of the railway line.

Cardiff City Council

We now run the planning of this area. Our Council Offices are here and we see this as a fitting development for Europe's newest Capital City. Having the new Welsh Assembly here is also a great boost for the area. Cardiff will benefit from more jobs and more revenue for the City Council.

Manager of Schott/ NEG TV tubes

We came here because of the excellent site close to many other hi-tech firms in south Wales. There is also a readily available skilled workforce.

▷ **Figure 4.26** The flats built in the 1960s

△ **Figure 4.27** Atlantic Wharf new private housing

Activities

2 Assuming your chosen role, what are your views about the way Cardiff Bay has been re-developed?

a What do you like (say why)?

b What do you dislike (say why)?

c Have you any suggestions on how you would like future development to take place?

Extension

- Have a class debate about Cardiff Bay. Put forward those views for and against the scheme and take a vote!
- Write an article for a magazine containing interviews with people for and against the scheme.
- Prepare a TV script (to last no more than two minutes) showing an audience all sides of the debate.

△ **Figure 4.28** New restaurant built on the Bay

▽ **Figure 4.30** The barrage turns the Bay into a freshwater lake

△ **Figure 4.29** St David's Health Spa and Hotel

How do Mexicans live in town and country?

Mexico is a large country of 1 958 201 sq km situated in North America. The total population is in excess of 93 million. The country consists of 31 states and one federal district where the capital Mexico City is situated. The country gained its independence from Spain in 1821 and as a result, the official language is Spanish, although Indian languages are spoken in some remote rural areas. The climate varies from a dry desert in the north to rainy tropical forest in the southeast. Most of the people live in the central **plateau**. The border with the USA stretches for over 3000 km and the country is at the narrowest point 225 km across from the Pacific Ocean to the Gulf of Mexico. Oil and tourism earns Mexico the most foreign income. Mexico is a developing country with a yearly income of about $4000 per head.

Two traditions are to be seen in the central square of Mexico City, the Zócalo (Figure 4.32).The Native American dancers can trace their heritage back to the time of the Aztecs over 600 years ago. The European influence can be seen in the old Spanish colonial buildings. The Spaniards conquered the Aztecs in 1519 and destroyed their large capital city of Tenochtitlan close to the Zócalo.

▽ **Figure 4.31** Indian dancers in the Zócalo

Most people in Mexico live in urban areas (74%) but there are still many Mexicans who live in rural areas. The population of Mexico is rising because the annual **birth rate** is higher than the **death rate**. Mexico's birth rate is 25 per thousand of the population per year and the death rate is only 5. The birth rate is often higher in rural areas where there is a long tradition of having large families. People are always needed to work in the fields and help with housework. The death rate is much lower than years ago because now there is better health, even in the countryside. 93% of Mexicans have access to health services.

△ **Figure 4.32** Ruins of Tenochtitlan, the Aztec capital

Activities

1 Figures 4.31 and 4.32 show the Indian origins of Mexico City as a settlement.
a Are the Indian dancers being exploited by the tourists or is it the other way round? What do you think?

b The ruins of Tenochtitlan are uncovered for residents and tourists to see. Why do you think there is less interest in the ruins compared to the Indian dancers? (Both pictures were taken at approximately the same time.)

2 Why is the official language of Mexico Spanish?

3 What will happen to the size of Mexico's population over the next few years? Explain your answer.

Tehuilotepec is a village 5 km north of Taxco, a town of 95 000 people in the Mexican state of Guerrero. Like many villages in Mexico many people have left to live and work in larger towns and cities. Why people leave their villages of birth has been explained by the ideas of **PUSH** and **PULL factors**. People are often pushed from a village because there are not enough jobs and living conditions are poor. A high birth rate means that villages have to support more people from the land. They are pulled to the towns and cities because they can get better jobs and feel that there is better lifestyle for their families. The pull is often about what people *think* is better. They may have seen a glamorous TV programme of life in the city or heard tales from a visitor who has moved there.

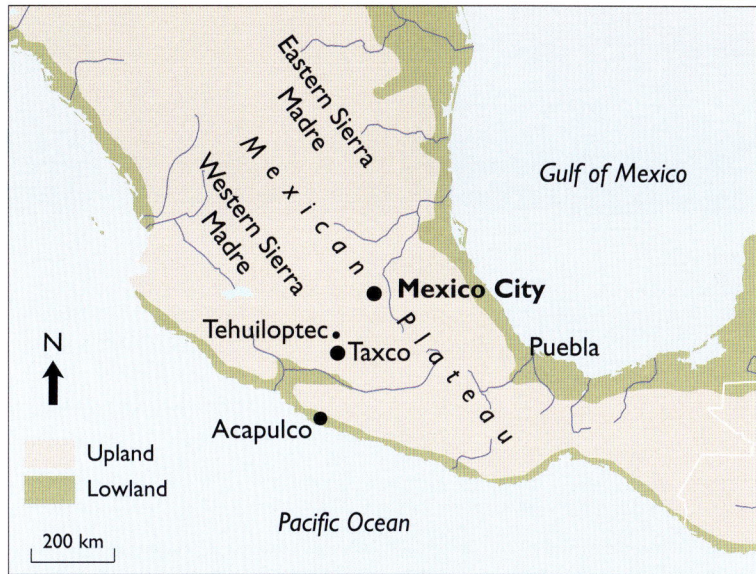

△ **Figure 4.33a** Location map for Mexico

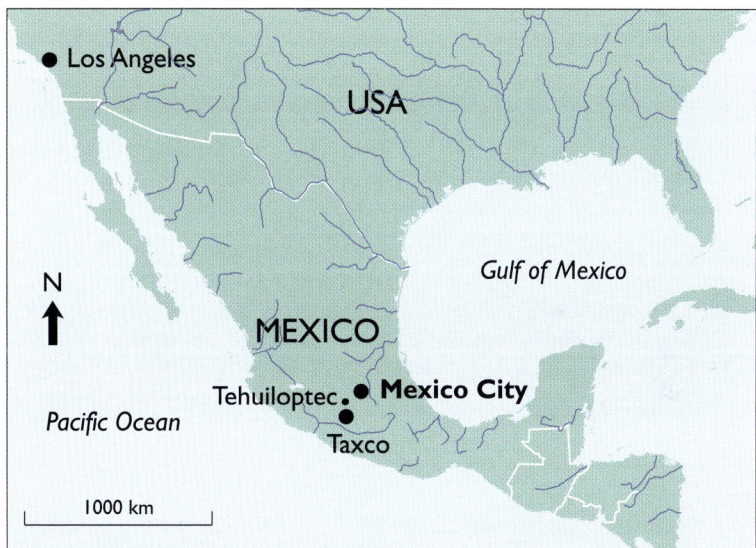

△ **Figure 4.33b** Map of Mexican plateau

Activities

4 Here are some reasons why people might leave the village. Put them into two types: PUSH and PULL.

- There's not enough food to go around.
- You can easily get an **informal job** such as shoe-shining.
- There is a chance to get a steady job in a factory.
- Schooling finishes after primary level.
- Jobs are few and mainly low paid or **subsistence**.
- Hospitals are nearby.
- Health care is basic.
- There is always something to do.
- There are good educational opportunities.
- It's not very exciting for young people.

◁ **Figure 4.34** Shining shoes is an informal job that any new migrant can take up

Village life versus the city?

△ **Figure 4.36** The village today

Until recently many people born in the villages of Mexico lived, married, raised families and died close to their village. Today there are more opportunities for young people who do not want to follow along the lines of their parents. The activity below highlights some of the choices now available.

◁ **Figure 4.35** 'The Village Scene', a print sold to the tourists

Activities

1 Figure 4.35 shows an image of life in the village. What are the people doing? Do you think it is an accurate view of life in a Mexican village today?

2 Figure 4.36 is on the edge of the village. What do you see to show that village life has changed compared to the print?

3 You are a young person from Tehuilotepec. You have just left school. Should you move or stay? Here are some options:

Stay in the village. Your parents need you to help on the small family-owned farm. You can travel daily to a job in the nearby town of Taxco catering for the tourists. The farm could be expanded to provide more food for the hotels and restaurants.

Move to Taxco, which has a population of 95 000. This town is an old historic silver mining centre with over 300 silver shops. With the improved highway links tourists arrive by coach from Mexico City, a three-hour drive away. You have the opportunity to sell tourist souvenirs such as the print (Figure 4.36), which is made by your family still living in the village.

Move to Mexico City. This city is one the largest in the world with over 19 million people. It's only a three-hour bus journey away. There are plenty of opportunities but you may have to take up an informal job such as shoe shining. Housing is difficult to find but you have an uncle who lives in a large shanty town on the northern edge of this sprawling city. The city has an enormous traffic problem that creates air pollution, made worse by polluted air being trapped at ground level. There are also problems with water supply and sewage.

a Which settlement would you want to live in and why? Explain why you would not want to live in the other two?

b Write a letter to a friend explaining your choice and what experiences you are having.

c Do you think you will be staying where you have chosen when you are older?

4 Look at Figures 4.37 to 4.41. Are these pictures how you imagined Mexico? Did any of them surprise you?

△ **Figure 4.37** Mexico City is over 2 200 metres above sea level and is bounded by mountains on three sides. It covers an area of over 2 000 sq km with more than a fifth of Mexico's entire population

△ **Figure 4.38** Mexico City shanty towns using the steep unwanted land

△ **Figure 4.39** Taxco is an historic town set in the upland plateau south west of Mexico city. Its buildings have now been protected. Tourism and the sale of silver jewellery is the main source of income

▽ **Figure 4.40** The Latin American Tower

△ **Figure 4.41** Traffic congestion is a major problem but there are plenty of taxis

△ **Figure 4.42** The old village shop, Hilderstone, 100 years ago

△ **Figure 4.43** The building today

Hilderstone is a village in the north Midlands of England in the county of Staffordshire (see pages 68–71). Like most villages it began hundreds of years ago. The village then was the centre of life for people who lived from farming or supplying local needs. Look at Figures 4.42 and 4.43. How has village life changed?

Recently, new houses have been built in this village (Figure 4.45) and there are others that could be bought and improved (Figure 4.44). There is concern in the village about whether more new houses should be built. A special Parish Council meeting has been called at the Village Hall to discuss the matter.

◁ **Figure 4.44** Housing that could be modernised

▽ **Figure 4.45** New housing

There are five main options:

A Build 20 starter homes on the edge of the village for local people on low income.

B Allow further development of old buildings and barns. They will need to fit in with the style of the village.

C Build 25 old people's bungalows for villagers and those living in nearby farms.

D Build 10 architect designed executive style homes. They will have double garages and 5 bedrooms. Some will have a swimming pool.

E Do nothing! The village should not expand.

Seven members of the public have turned up and want to air their views. They are:

- A young married couple with two toddlers. There is only one wage earner in the family on a low income.
- A recently retired farmer who wants to move into the village after selling his local farm.
- The owner of the pub 'The Roebuck'.
- A manager of a pottery firm in Stoke-on-Trent who is looking for a large family house and would like to live in the 'country'.
- A farmer who owns land on the edge of the village.
- A local conservationist who often gives talks in the Village Hall.
- An unemployed labourer, living in the village.

△ **Figure 4.46** Old people's bungalows

Target task

1 Choose two of the options for Hilderstone. Suggest one advantage and one disadvantage for each of these options.

2 Choose two of the people in the list. Match them to the option you think they would prefer. Explain why.

3 What do you think should happen, and why?

Extension task

4 Set out in a table the social, economic and environmental advantages and disadvantages of each of the five options.

5 Match each of the people in the list with the option you think they would prefer. Justify your selections.

6 Which are your preferred and least preferred options? Explain why.

△ **Figure 4.47** The Roebuck pub

▽ **Figure 4.48** The Village Hall – a centre for local events

Assessment tasks

Report writing

Choose one of the case studies mentioned in this chapter, Hilderstone, Mexico City or Cardiff. You should prepare three different reports. You should keep them brief (no more than two sides):
• One should be in the form of a holiday brochure.
• The second should be for inclusion in a local government report.
• The third should be for children in a primary school.
For each you should include text, at least one image and some statistics.

Remember to write your report according to the needs of the audience. Primary school children will need less words and clearer information than a technical report for local government officials. Holiday brochures will need to be attractive.

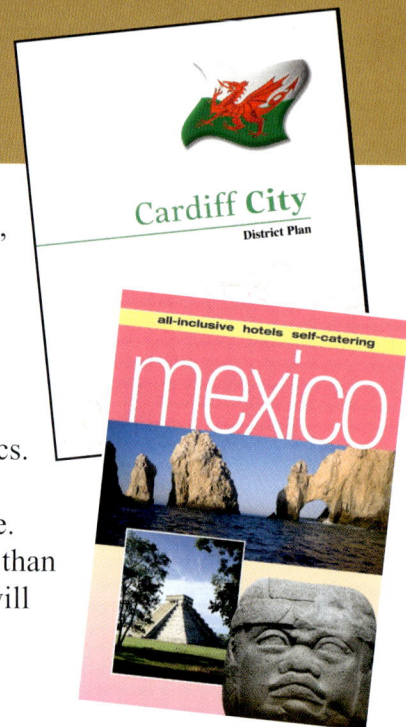

Cardiff City District Plan

all-inclusive hotels self-catering

mexico

Designing a website

ICT activities

1 Draw a flow diagram to show the structure of a website for the settlement you live in. Figure 4.49 gives an example of how you might start it. Think of the things that might interest people of all ages. What information would you want to know?

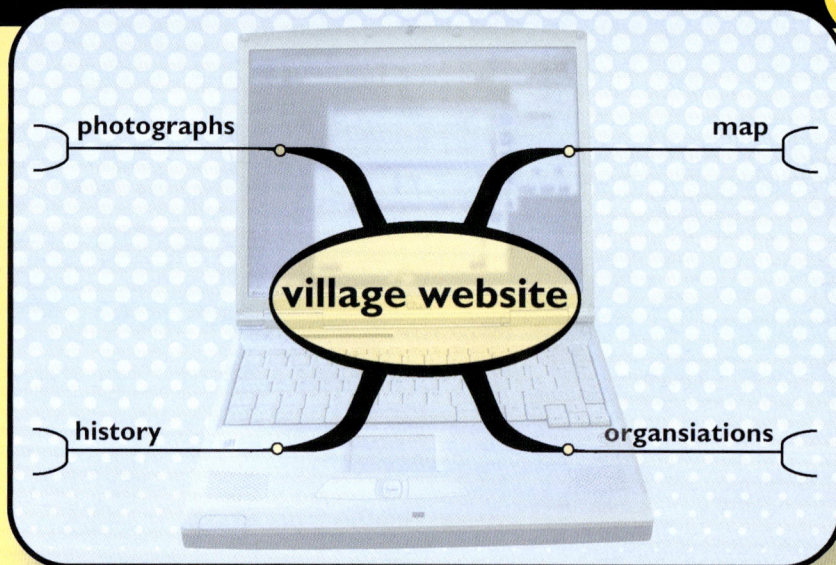

photographs   map

village website

history   organsiations

△ **Figure 4.49** Flow diagram for a village website

2 Here are some websites that may give you some ideas. There are several sites for villages. Try finding out about a village you know – type its name into a search engine such as www.google.co.uk.

Web address	What it shows
www.hilderstonevil.freeuk.com/	Shows information about Hilderstone
www.old-maps.co.uk/	Finds an old map of your village
www.ordnancesurvey.co.uk/getamap	Finds an up-to-date map
www.local-transport.dtlr.gov.uk/ruralbuses/	Contains information on the government plans for buses including rural areas

Review

Using photographs to review your work

One way of finding out whether you have understood this chapter is by selecting two photographs that you *think* are different; for example, one could show an urban area, the other, a rural area. Use the compass rose (Figure 4.50) to help you analyse them.

〉 The compass rose

1 Sketch the image.
2 Place this in the middle of a sheet of plain paper. Try and describe each photograph under the following headings:

△ **Figure 4.50**

N	**North** stands for Natural. Describe what is natural about the photograph, e.g. vegetation, clouds
S	**South** stands for Social. Describe how people's lives are affected, e.g. where do children play?
E	**East** stands for Economic. Describe how money is made, e.g. what jobs are shown?
W	**West** stands for Who Decides? Describe the decisions made by people that have influenced the photograph, e.g. why a building is here?

Does the compass rose help you to find any similarities? Now choose another two photographs that you think are different – you could use images from this chapter or select others from books or the Internet.

Think about this

Do big settlements have a future? Are they **sustainable**? All settlements will have an 'ecological footprint'. This is the land needed to feed them, supply them with timber and reabsorb their carbon dioxide emissions. Big settlements have giant 'footprints'. London has 12% of Britain's population and has a 'footprint' of 21 million hectares, which is 125 times the area of London! Think of ways in which a settlement's 'footprint' can be made smaller, e.g. recycle products, plant more trees. Why do you think it is a good idea to make these 'ecological footprints' smaller?

5 Ecosystems

How are you linked to a blade of grass?

Ecosystems are all about survival. Every animal, plant and insect has its part to play in the continued survival of the whole system. Think about these parts of an ecosystem. Who eats whom? There are some obvious answers, but some of the answers are less clear.

Most likely to eat or be eaten?

▷ **Figure 5.1** Who eats whom?

What are ecosystems?

Ecosystems are made up of living organisms (**biotic**) and non-living things (**abiotic**). All these parts, or **components**, of ecosystems are linked with each other in a complex way. Understanding the way different components are linked is an important part of studying ecosystems.

Ecosystems are not all the same size. They vary in scale from those which cover huge areas of the planet, like a rainforest, to those that may be less than 1 m², like a garden pond.

Whatever their size they will have biotic and abiotic components dependent on each other for their survival. It is the complexity of these links that makes it very difficult to predict how ecosystems will be affected by changes caused by humans or natural events. Often such changes result in long lasting and irreversible alterations in the ecosystem, like species extinction.

What has been said about the natural world

'We mention nature and forget ourselves in it: we ourselves are nature'

Friedrich Nietzsche (1844–1900)

'Nature is the symbol of the spirit'

Emerson (1803–1882)

'Extinct is forever'

Friends of Animals

'He who plants a tree, plants hope'

Lucy Larcom (1826–1893) U.S. poet

△ **Figure 5.2**

The battle for survival between the organisms of an ecosystem creates a natural environment full of competition. Species are said to **evolve** this way, with only the best suited organisms able to survive by adapting to the ecosystem's environment.

Human beings have evolved to a stage where it is possible for us to learn how to survive in a wide variety of conditions found in many different ecosystems. So humans have developed to use the resources in ecosystems in different ways to other living things.

Figures 5.3 and 5.4 show the same ecosystem being used by two different species of organism, leaf cutter ants and humans. They are both doing the same thing: taking natural resources from an area of tropical rainforest in Brazil. However, the effects on the ecosystem they are exploiting are very different.

During this chapter you will look at examples of ecosystems and where the natural environments have been put under threat by humans using resources for their own development. When you see these examples, think about the balance needed to ensure that ecosystems can be used but not destroyed by people.

△ **Figure 5.3** Leaf cutter ants

▷ **Figure 5.4** Humans taking resources from the rainforest, Brazil

Activities

1. Describe one of the links you made in the Who eats whom? diagram (Figure 5.1).

2. What would happen to the other organisms if one was removed?

3. Is one organism more important than another? Explain your answer.

ICT links

During this chapter we will look at a number of examples of ecosystems. There are many great resources on the Internet. Try to do your own research using the following sites as a starting point:

www.nationalgeographic.com/earthpulse

http://mbgnet.mobot.org/index.htm

What do you already know about ecosystems?

You will be surprised at how much you already know about the way ecosystems work and how the parts, or components, working inside them are linked to each other.

Figure 5.5 shows all the main components of a large scale ecosystem. It may be easy to say which parts are linked, but it is your job to *explain the links between components*.

Activities

1 Complete the concept map, Figure 5.5.
The key to a good concept map is detail. When you explain the links between components, always try to give examples to help expand the detail in your links.

2 What else could you add? There are only six components with ideas in this concept map, so why not add your own components as you think of more important parts of ecosystems. How would humans fit in?

Systems Thinking

Looking at parts of the natural world as if they were a system can help us simplify some of the complex relationships between living organisms and their environment.
Like any other natural system, ecosystems can be broken down into key parts, all of which contribute to what we see.

- Inputs
- Flows
- Processes
- Stores
- Outputs
- Feedback

▽ **Figure 5.5** Concept map

essential for survival

Animals — **Water**

Sun **Plants**

Soil plants use nutrients in soil to grow **Rocks**

What is the point of studying ecosystems?

During the rest of this chapter, you will need to ask yourself some questions about how humans fit into ecosystems.

Some say that the natural world is there for us to use as we see fit. Other groups of people hold different views; they think that we should take care of our planet for all the generations to come.

The idea of **sustainable development** is a way of looking at protecting our planet's resources into the future. It means development that does not damage the environment for the future. For example, if you chop a tree down you would have to plant two in its place. This kind of management is more easily done with some resources than with others.

Generating energy for today's hi-tech societies means using oil, gas, coal and nuclear power. These are all impossible to use sustainably, and can generate big pollution problems as well as electricty. Using alternative energy sources like solar power, wind and hydroelectric generators is one solution that takes the pressure away from some ecosystems.

So, as you work through this chapter, keep these ideas in mind because you are the ecosystem managers of the future.

Activities

From the list of words below pick the odd ones out and explain why you think they do not belong in the list. You may have to argue your case, so make the explanations good!

Exploitation	Power
Ecosystems	Destruction
Energy	Extinction
Money	Research
Politics	Development

ICT links

To find out more about alternative energy supplies and sustainable development, follow these links:

The Centre for Sustainable Energy
www.cse.org.uk

The World Wildlife Fund, Sustainable Development site
www.wwflearning.co.uk/ourworld

Structure your investigation of ecosystems

At the end of this chapter there is a review page. It has a table showing the key learning goals for this topic. Before you go any further why not have a look at these goals and the steps leading up to them (on page 103). They will help you work through this chapter much more effectively.

▷ **Figure 5.6** Examples of sustainable use of resources: (i) wind farm, (ii) replanting forests, (iii) re-seeding a flower meadow, (iv) solar panels on a house

What is the global distribution of ecosystems like?

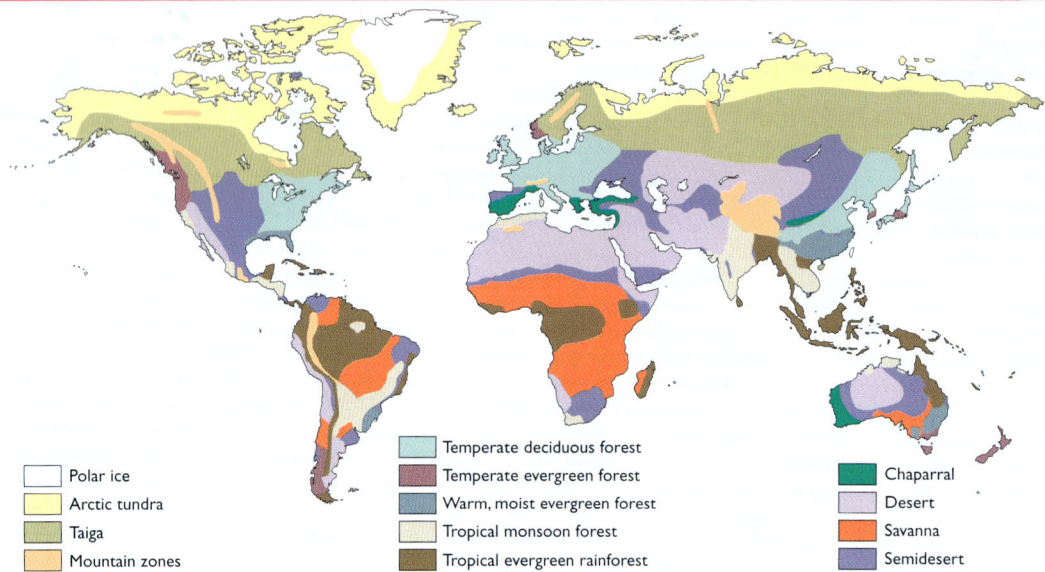

Key:
- Polar ice
- Arctic tundra
- Taiga
- Mountain zones
- Temperate deciduous forest
- Temperate evergreen forest
- Warm, moist evergreen forest
- Tropical monsoon forest
- Tropical evergreen rainforest
- Chaparral
- Desert
- Savanna
- Semidesert

△ **Figure 5.7** Distribution map

The distribution of all ecosystems is controlled by three simple ingredients of life on earth: light, heat and moisture.

These three factors create the conditions for all life on earth and so control where different ecosystems are found. When you look at the map (Figure 5.7) you can see that there are many parts of the world with the same kind of ecosystem. These large-scale ecosystems are called **biomes**.

These biomes have developed in areas with similar conditions for life. As climate plays such a large part in creating these conditions, there is a strong link between climate and distribution of the world's biomes.

The energy from the Sun provides the Earth with light and heat. This in turn creates an atmosphere filled with moisture evaporated from the Earth's surface. However, as not all the Earth gets an equal share of the Sun's energy, climate varies from place to place and helps cause the distribution of biomes seen in Figure 5.7.

The Sun's energy is concentrated at the Equator and weaker at the poles. Here the curve of the Earth's surface means a larger area needs to be heated by the same amount of energy as at the Equator. With the changing seasons, the poles also tilt toward and away from the Sun much more than the Equator. This means the length of daylight varies greatly, and with it the growing season for all plants.

KEY
A: Passage through the atmosphere
B: Angle of incidence
C: Area heated

◁ **Figure 5.8** How the Sun's rays are incident on the globe

What other factors affect the distribution of ecosystems?

There are other factors that influence the distribution of different ecosystems, but not on such a large scale as global climate zones.

The wide variety of plants and animals in the world reflects their ability to evolve and adapt to their environment.

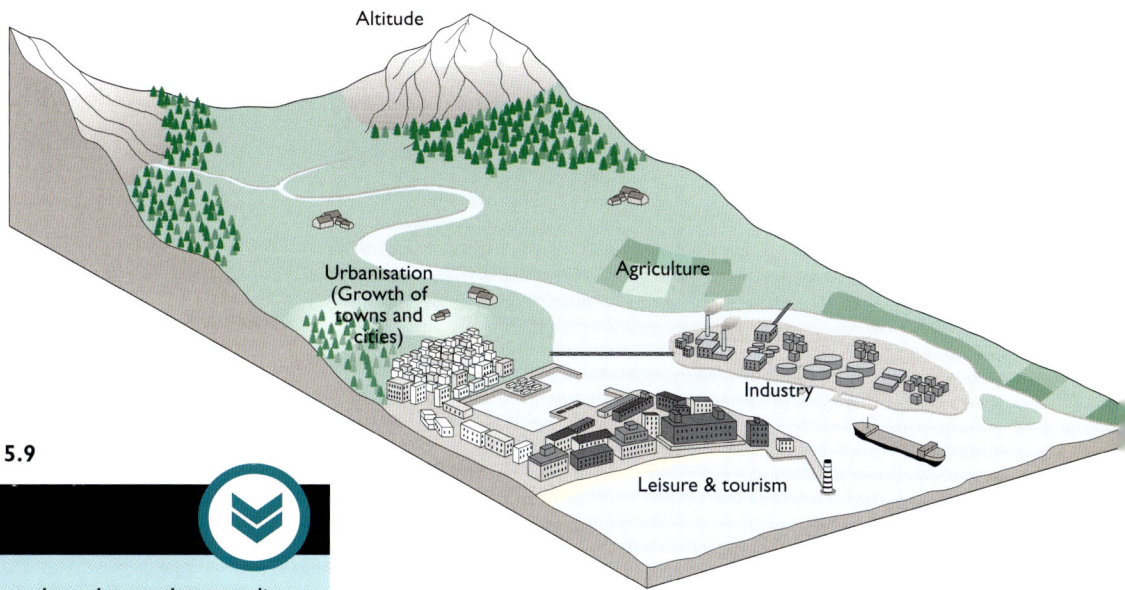

▷ **Figure 5.9**

Altitude

Urbanisation (Growth of towns and cities)

Agriculture

Industry

Leisure & tourism

Activities

As animal and plant species evolve, they adapt to live in different climatic conditions. This means it is possible to have closely related species living in very different ecosystems. But what are the reasons for these adaptations?

Now you are going to think about evolution and adaptation in a slightly different way!

1 Use your imagination and the characteristics listed in the box below to help you make the following animals:

A herbivore living in a hot climate
A carnivore living in a very cold climate

A good way to begin is to draw your new animal. Label the adaptations it has for life in its ecosystem. You will need to add more characteristics to the ones you have been given to do a really good job creating your new animal.

Surviving the climate	Finding food	Avoiding trouble
Thick fur coat	Fast runner	Camouflage
Active at night	Sharp claws and teeth	Smells bad
Doesn't need much water	Good sense of smell	Lives in large groups

The way both plants and animals have adapted to live in specific ecosystems is amazing. However, it can be a real disadvantage when conditions in the ecosystem change too quickly for them to adapt to a new environment. When this happens they can become extinct.

There are few ecosystems on earth that have not been affected in some way by human activity. People don't always get the balance right! There are plenty of examples of species of plants and animals which have become extinct due to the actions of humans.

ICT Activity

Using a search engine like:
www.google.com
www.lycos.com
www.yahoo.com

find out more about these extinct species:

• Mammoth
• Dodo
• Giant Sloth

How are ecosystems organised?

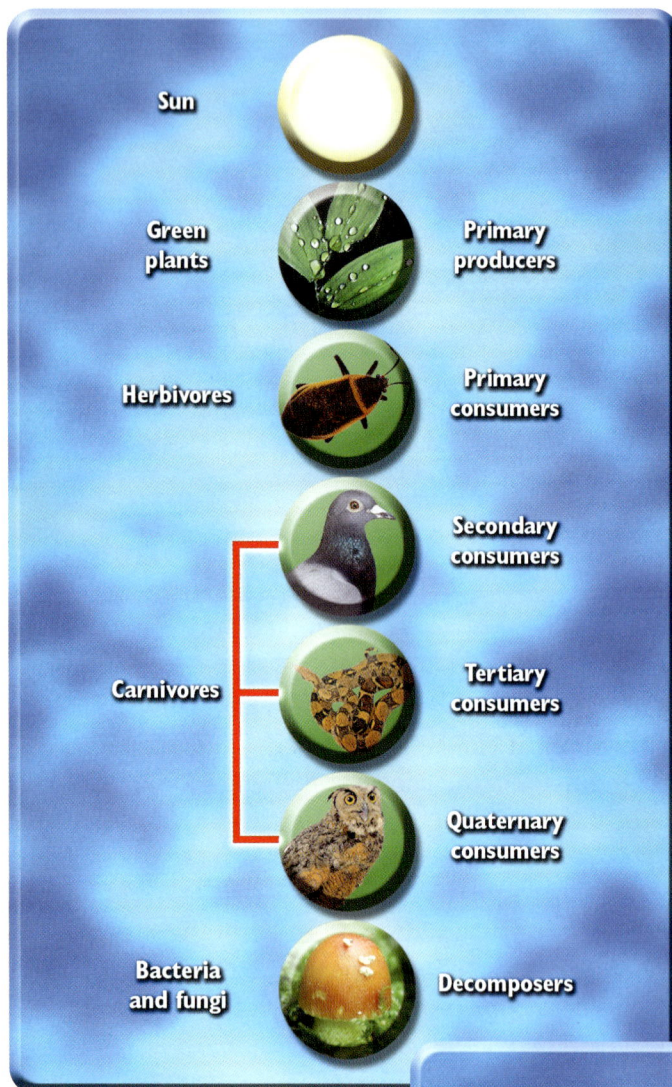

△ **Figure 5.10** A food chain

The individual links between components of an ecosystem are normally quite simple. However, when we look at the whole ecosystem, the number and nature of the links can be very complex. To simplify our understanding of how ecosystems work, we look at the way energy moves through them. We can also divide the components into different categories depending on the part they play in the ecosystem.

As energy passes up the **food chain**, it is used by fewer organisms. The organisms at the top of the food chain are usually more complex and able to prey on more than one species from a lower level.

The links between different organisms in a food chain can be shown on a wider scale through a food web (e.g. Figure 5.11). Here you can see that there are often many links between organisms.

The food web also shows how ecosystems are dependant on all the components in the food chain to remain in balance with their environments. When the food web is damaged, problems will occur.

A simple food web

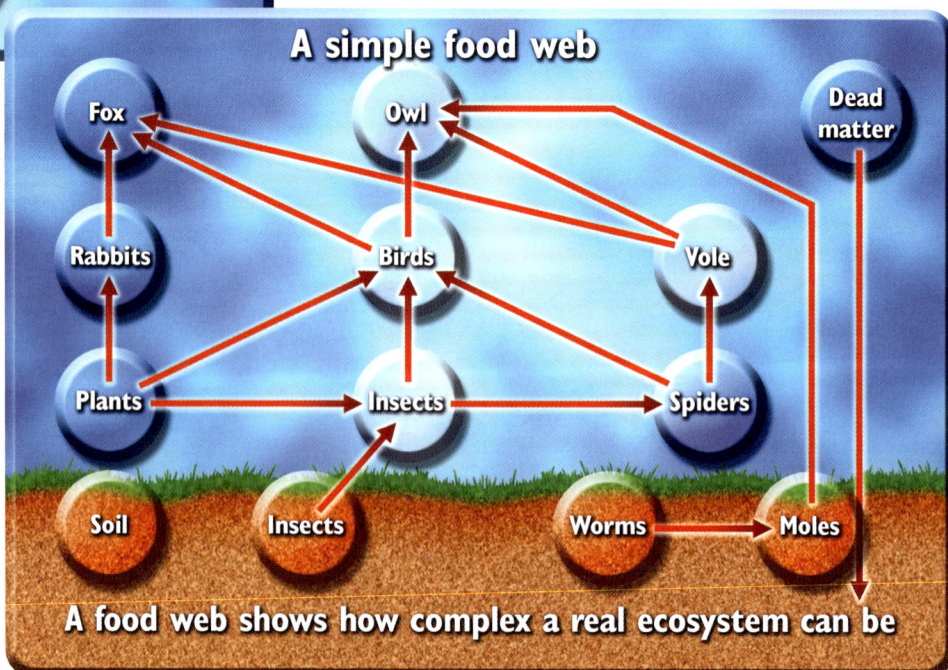

A food web shows how complex a real ecosystem can be

▷ **Figure 5.11** A simple food web

Cyanide Spill Floods into Danube

Nick Thorpe in Budapest, Monday 14 February, 2000 for the *Guardian*

A flood of toxic cyanide 40 km long washed down the River Tisza in northern Yugoslavia yesterday accompanied by a tide of dead fish, spreading the poison further downstream into the Danube and adding to the toll of environmental destruction across central Europe. The spill originated in northern Romania, where a dam at the Baia Mare gold mine overflowed and caused cyanide to flow into streams and rivers.

'If I am pessimistic, I would say that life in the River Tisza will never recover from this,' said Elemer Szalma, a biologist from Hungary. 'If I am optimistic, I would say it will take 10 to 12 years to recover.'

On Friday fishermen stretched nets between barges in Szeged, and removed five tonnes of dead fish. So far 300 tonnes have been removed from the Tisza.

The Mayor of Senta in Yugoslavia estimated that 80% of the fish in the Tisza, where it flows through Senta, were dead. Mr Szalma said:

'Even if we reintroduce fish into the river, they will die of starvation because the food chain has been completely broken.'

Yugoslav officials added their voices to Hungarian and Romanian demands for compensation from the Australian company that owns a controlling share in the gold mine. But company representatives have played down the disaster.

'There is no doubt that a significant amount of water overflowed the dam,' said the chairman of the company, 'but this is not an ecological catastrophe.'

△ **Figure 5.12**

Activities

Foundation

1 List all the parts of a single food chain shown in the food web diagram (Figure 5.11).

2 Could any parts of your food chain survive without the layer below it?

3 Look again at the whole food web. How could it be damaged by human activity?

Target

4 Make two lists from the new report.
a Facts about the incident.
b Opinions of people involved.

5 From your lists, pick one fact and one opinion that you think is most shocking or important. See what other people have chosen.

6 What will be the long term impacts on the people who depend on the river for their living?

Extension

7 Why do think ecosystems are so vulnerable to damage from human activities?

8 What could the authorities do to prevent this type of problem happening again?

9 How can we define the term 'Ecological Catastrophe?'

Tropical Rainforests: A biome under threat

A warm, moist wind blew from the south down a valley lost in green folds of undulating rainforest. Carried on the wind were thick, earthy smells of sweet-scented plants and decomposing leaf litter that carpeted the jungle floor. This enticing fragrance conveyed the first hints of what life would be like within the 120 000 square miles of rainforest than separated me from the coast by more than 400 miles.

◁ **Figure 5.13** From *Stranger In The Forest* by Eric Hansen

This description of part of the Amazon Rainforest just hints at the type of place it is. Rainforest biomes cover large areas of the world, occurring in a belt of latitude roughly between 10° N and 10° S of the Equator. This position means they share a very similar climate.

▷ **Figure 5.14** World map of forests

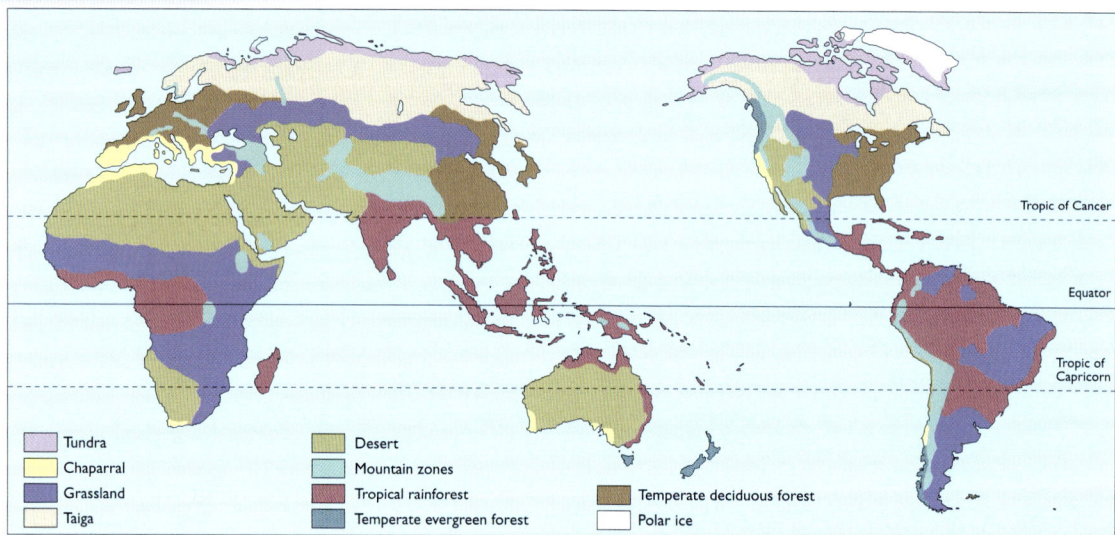

Legend:
- Tundra
- Chaparral
- Grassland
- Taiga
- Desert
- Mountain zones
- Tropical rainforest
- Temperate evergreen forest
- Temperate deciduous forest
- Polar ice

Tropic of Cancer
Equator
Tropic of Capricorn

CLIMATE

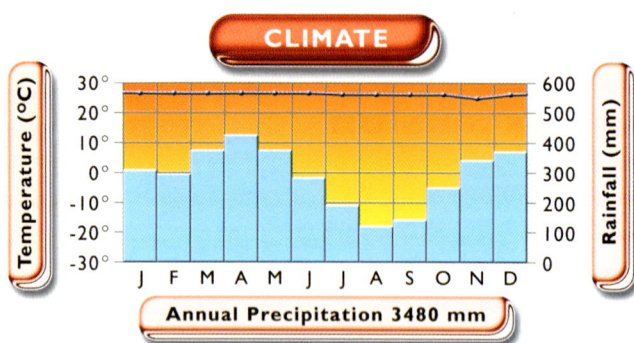

Annual Precipitation 3480 mm

△ **Figure 5.15** Climate graph for the rainforest region

◁ **Figure 5.16**

As you can see from the climate graph (Figure 5.15), the temperature in tropical rainforest areas does not vary much throughout the year. It is always hot! As for rainfall, there is plenty of it, with most days having downpours of heavy rain. The hot, moist conditions create convection clouds during the day which burst by the late afternoon.

With this hot, wet climate, it is not surprising that **vegetation** grows quickly. It also grows big! Imagine trees ten times the height of your house, and leaves large enough to cover your bed. The vegetation of the tropical rainforests is an amazing site!

These plants grow all year round as there are no real seasons in these areas. It doesn't stop there either, as this vegetation supports an amazing variety of animals and insects like no other biome on earth.

The tropical rainforests of the world are an incredibly important part of the global ecosystem.

EMERGENT LAYER
• The tallest trees called Emergents. Can be 70 metres high.
• The most sunlight.
• Eagles, monkeys and bats live here.

CANOPY LAYER
• The roof of the rainforest, about 30–40 metres high.
• Many trees produce fruit all year round in this layer.
• Most wildlife lives in this layer of the forest: birds, monkeys, tree frogs, and snakes.

UNDERSTOREY LAYER
• Plants must grow large leaves to catch what sunlight they can.
• Plants grow to 4 or 5 metres unless they get more light.
• Animals include big cats like jaguar and leopards, also many species of insect.

FOREST FLOOR
• Very little grows on the forest floor as it is so dark.
• Things that fall from the upper layers decay very quickly in the hot, damp conditions of this layer. Many species of insect and fungus speeds up this process.
• Ground living animals like giant anteaters, Capybara and wild boars live here.

Activities

1 Rainforests are full of life. Why do things grow so big and so quickly?

2 Why is most life in the rainforest concentrated in the canopy layer?

3 What things would you like and what would you dislike about living in a rainforest?

Facts and figures

• Trees in tropical rainforests can grow to enormous heights, sometimes reaching up to 60 m.
• The world's rainforests are home to at least 50% of all the world's species.
• Over 600 new species of beetle have been discovered in studies of a single species of tree.

Structure of the ecosystem

Although the forest may look a chaotic unorganised place, it actually has a well defined structure. As you can see from the diagram opposite (Figure 5.17), different layers exist at different levels above the forest floor. This structure depends on the way the plants of the forest battle with each other to gain the most sunlight.

The canopy layer is the most rich in life; it intercepts most of the sun's energy and is a dense mass of branches and leaves. It is also where fruit is most common and so is where much of the forest's animal and bird life is to be found. The nearer you get to the forest floor, the darker it gets. The floor is where much of the dead and decaying material ends up. It is also home to organisms like fungi and termites which do the job of recycling these valuable resources.

Nutrient cycling

With the warm, moist conditions created by the climate, both growth and decay of plants is faster in tropical rainforests than in other biomes. The nutrients which come from decomposing organisms are used by plants as food. They get most of these nutrients from this source, as the soil in these areas is often low on nutrients.

How are the rainforests under threat?

Reasons for Exploitation

With the climate of the tropical rainforests creating ideal growing conditions for vegetation, it is no surprise that these areas contain some of the richest supplies of natural resources in the world. Also, the climate means that they are relatively easy to exploit, and with growing markets for timber, oil, paper and other products of the rainforest, it is easy to see the pressure from developers that some of these areas are under.

Nobody knows when the demand for resources from the rainforest will slow or stop, but we do know that there is only a limited amount of forest left which has not yet been destroyed by human exploitation of these areas.

On these two pages you can find out more about some of the pressures on the world's tropical forests.

▷ **Figure 5.18a** Timber extraction for hardwood furniture in Balikpapar, Indonesia

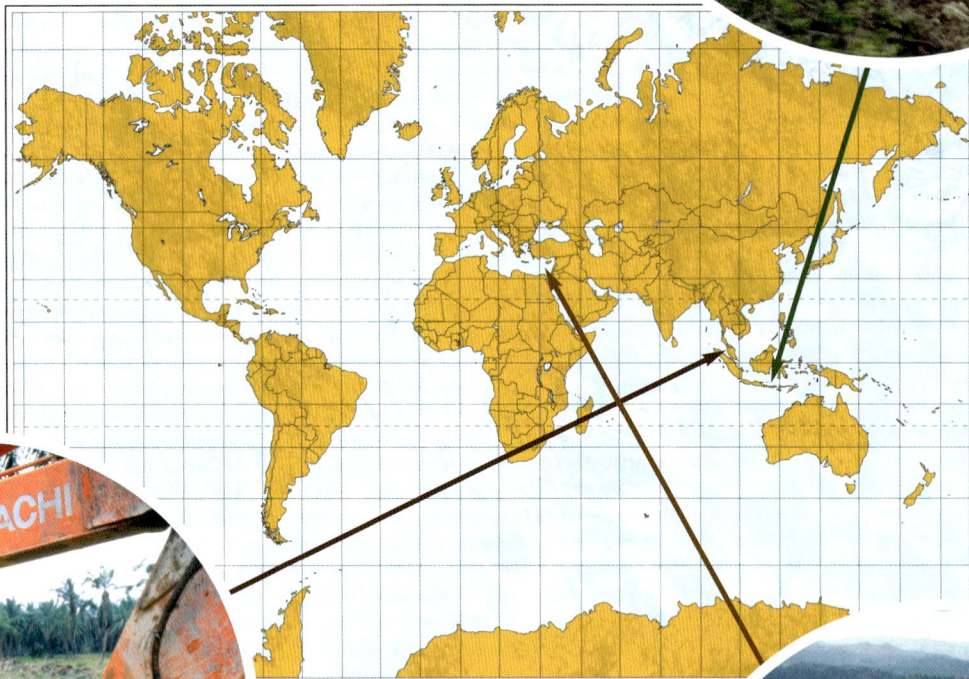

△ **Figure 5.18**

◁ **Figure 5.18b** Extensive oil palm plantation in Malaysia

▷ **Figure 5.18c** Skomiotissa mine, second largest copper mine in Cyprus

Medical products and research

Wild yams from the rainforests of Mexico and Guatemala have been very important in the production of the active ingredients in birth control pills, diosgenin and cortisone.

ICT links

www.ran.org/info_center/factsheets/05f.html

www.rainforestweb.org/Rainforest_Destruction/Mining/

Timber extraction

Rapidly increasing levels of wood and paper consumption is one of the primary factors driving global deforestation. More than half of the world's timber and nearly three-quarters of world's paper is consumed by a mere 22% of the world's population – those living in the United States, Europe and Japan.

Most tropical timber imported into the US is in the form of plywood and panelling, produced in Indonesia, Malaysia, Taiwan and other Southeast Asian countries.

Oil palm plantations

The oil palm is a good example of how large-scale, commercial agricultural plantations are the driving force behind deforestation.

Malaysia is the world's top producer and exporter of palm oil, generating 50% of the global output, of which 85% is exported.

47% of Malaysia is forested, with a total forest area of approximately 37 million acres. Since the mid-1990s, Malaysia's deforestation rate has been over 2.4% annually.

Cattle ranching

Cattle ranching is a major cause of rainforest destruction in Central and South America. Ranchers slash and burn rainforests to grow grass pasture for cattle. Once the cattle have grazed sufficiently, they are slaughtered and exported to industrialised countries, including the United States, to be made into fast food hamburgers and frozen meat products. It has been estimated that for every quarter pound hamburger made from rainforest cattle, 55 square feet of rainforest was cleared.

Mining & quarrying

Mining, particularly gold mining, is an increasing threat to the world's rainforests and to forest communities. In large-scale gold mining operations, enormous pits are dug out of the land; dynamite is often used to blast holes in the ground; ore is sprayed with cyanide solution to leach out the gold.

In Ghana, the only remaining rainforest is located in the same area as the country's gold mines. The mines are destroying forests, communities, and traditional livelihoods. Approximately 54 000 acres of rainforest are destroyed in Ghana each year, primarily to make way for transnational corporations seeking to profit from gold. As a result, entire communities have been forced to relocate.

Activities

1. Write a list of five facts that you have learnt about the destruction of the rainforests from the reports on these pages.

2. Would you use a medical product if it were developed at a cost to the rainforest? Explain your answer.

3. As **consumers**, what power do you think we have to influence the multinational companies working in the rainforests?

4. As an MEDC do we have the right to stop LEDC developing their natural resources to make progress for their people?

How can ecosystems be managed?

There are hardly any ecosystems left on earth that have not been affected in some way by people.

Often when humans have come into contact with the natural environment their first instinct is to see what they can get from it and how they can benefit from its resources. These resources may be obvious, like timber from rainforest trees, or less obvious, like an underground oil reserve. They can even be resources which we cannot actually touch, like a spectacular view from the top of a hill, or the experience of walking on a beach.

Only in the last few years have people really begun to think about the damage they may be causing to the ecosystem. Today it is much more likely that people who wish to use the resources of an ecosystem will try to limit the damage done by some form of management.

Aims of management

Managing ecosystems is all about balance – a balance between helping those who want to use the resources and making sure that any damage done is not permanent and is kept to a minimum. Management should be targeted at protecting and conserving the ecosystem.

It is often the case that ecosystems provide many different resources to different types of user. For management to be successful it has to look at areas where different user groups may conflict in the demands they put on the ecosystem. Where **conflict** occurs, it is important to try and resolve it through management.

Conflict resolution

Conflict arises when more than one group wants the same resource at the same time. Think about your family and friends. How many times have they clashed over using the TV, computer, telephone or bathroom?

Timing is one way to resolve these conflicts, with different groups allowed access to the same resource at different times.
Zoning is another way, where the resource is split into separate areas and each group just stays within their space.

There are other ways of managing conflict, can you think of any?

What is management?

We use the word management a lot in geography, often in connection with many different topics, but what does it actually mean?

In most situations it is used to describe **the control of a variety of factors to achieve a specific purpose.**

What is sustainable management?

To use the resources of an ecosystem so that it is maintained.

▽ **Figure 5.19** Conservation area

Buffer zone management

This way of controlling access to sensitive areas can be used in many different situations and at a variety of scales where ecological damage needs to be prevented.

It has been put into practice all around the world, with a famous example in the Rainforest of the Korup National Park in the West African country of Cameroon.

The Core Area
No access to the core is permitted. There is as little human contact with the core as possible. Some scientific monitoring may be allowed.

The Transition Zone
While there are no restriction on the number or type of activity that happen here, all potential threats are monitored closely.

△ **Figure 5.20a**

△ **Figure 5.20b**

The Buffer Zone
Limited access is permitted for certain activities; for example, agriculture, recreation, education and scientific research. People's movements are tightly controlled with penalties for those who ignore the rules.

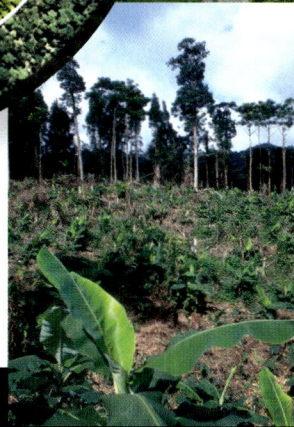

◁ **Figure 5.20c**

Activities

1 As a class, come up with a definition for the word *resources*.

2 Test our definition of management. Do football or supermarket managers have an aim in the way they control their resources?

3 What form of conflict resolution is buffer zone management?

4 Take a piece of plain A4 paper and draw a border around it, 1 cm in from the edge. The rectangle you have now made represents an area of rainforest where 4 cm represent 1 km (roughly). You are responsible for its management.

There are some conditions though:
- 25% needs to remain untouched and conserved as primary forest.

In the rest you must fit:
- two hotels
- an area for visitors to trek in the forest
- five villages with farmland to supply the hotels with food
- any roads or tracks for transport.

The aim is to manage the area to create minimum conflict between the users and resources. Compare your plan with others in your class.

How does management work on a small scale?

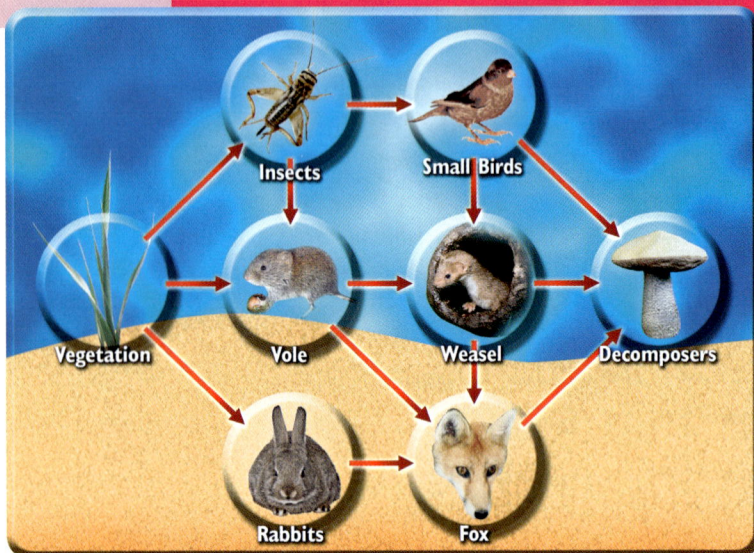

△ **Figure 5.21** The sand dune food web

Coastal environments contain a variety of unique ecosystems. In the UK one of the most interesting are sand dune ecosystems.

The sand dunes found along the north-east coast of England are an important part of the country's ecological diversity.

Druridge Bay is a part of this impressive coastline. It stretches from Hauxley in the north to Creswell in the south. The sand dune system is protected by UK, European and International laws which aim to prevent damage to the plants, insects, birds, animals and geology of the area.

The country park is popular with visitors all year round, while the nature reserves, which are less heavily managed, also attract people wishing to experience the sand dunes.

There are many examples of good ecosystem management at Druridge Bay, but what are they and how is management carried out?

The visitor centre has a large car and coach park. It also provides toilet facilities and has a small café. The centre acts as an information point with a number of boards and leaflets informing the public of sand dune ecology and recent management activities.

The dunes in front of the visitor centre have a high level of management. Paths direct people to the sea, with fences preventing access to sensitive areas.

Further south the dunes near Creswell have a much lower level of management. They are less well used and only have a number of informal paths created by visitors. There are no restrictions on acces for people.

© Crown copyright 2002

◁ **Figure 5.22** Map of Druridge Bay

△ **Figure 5.23**

People visit the sand dunes for different reasons. The diagram above shows some of them. It also shows the potential problems, or threats, these activities may cause to the ecosystem.

One of the best ways to prevent these problems is to control where people go when they are in the dunes. Fences, steps and board walks direct visitors along certain paths where management can be targeted.

Management to protect the dunes

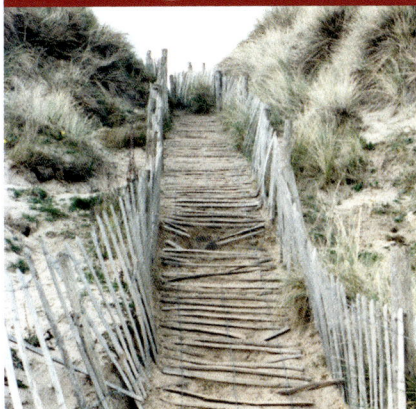

△ **Figure 5.24** Fences and steps

△ **Figure 5.25** Board walks and paths

△ **Figure 5.26** Visitor centre and car park

Activities

Foundation

1 Look at Figure 5.21. How could visitors to the dunes damage the links between these parts of the **food web**?

2 Make a copy of Figure 5.23.

a Add any more threats you can think of.

b Around the outside add your ideas about how to prevent these problems from happening.

Target

3 Look back at the definition of management on page 96 and list the aims of management at Druridge Bay.

4 The dunes are protected by law. Find out what a 'Site of Special Scientific Interest (SSSI)' is in the UK.

5 Board walks and steps are two ways of stopping path erosion. Describe one more.

Extension

6 Is a sign saying 'Do not pick wild flowers' management?

7 A small species of snail found only in this part of the country is one reason for the SSSI protection of the dunes. Is it right to stop people enjoying some of the area because of something the size of a five pence piece?

Coral reefs have been described as 'Essential Life Support Systems' (World Conservation Strategy 1980) necessary for food production, health and other aspects of human survival and sustainable development.

ICT links

www.fisheyeview.com

www.cyberlearn.com

◁ **Figure 5.27** World distribution of coral reefs

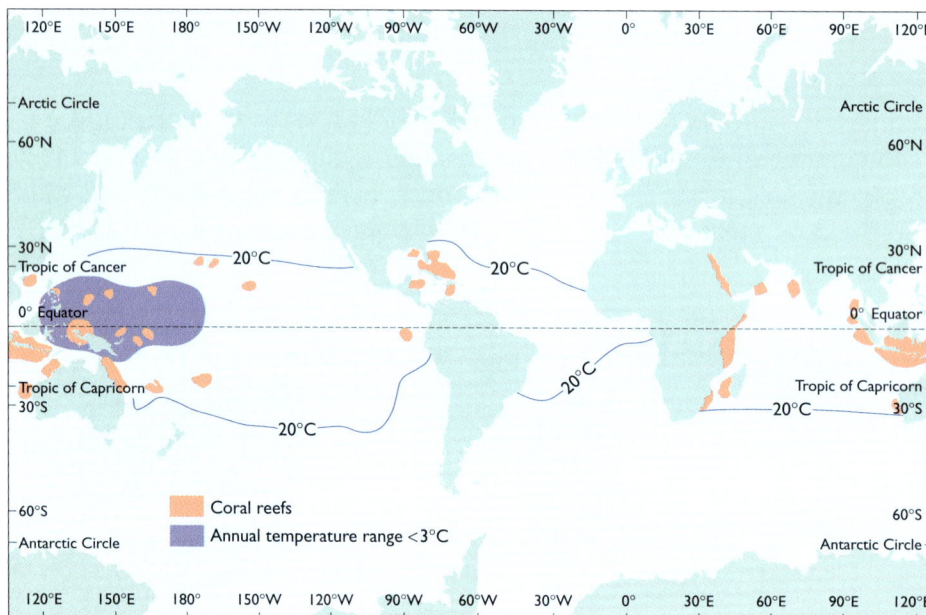

△ **Figure 5.28** Coral reefs

What are reefs like?

Coral reefs can be found in the tropical water of over 100 countries which generally lie between the tropics of Cancer and Capricorn 23.5°N and S of the Equator. It has been estimated that the total global area of live coral is nearly 600 000 km².

Coral is made from the skeletons of tiny organisms called polyps. They only grow in warm, clear, shallow sea water. Some types of coral grow quite quickly, but most grows very slowly to form ancient reefs. These are home to thousands of species of marine life.

Coral reefs are very old, rare and sensitive ecosystems that cannot survive great changes in their environmental conditions.

Pressure on reefs

As coral grows in warm, clear seas, it is no surprise that these reefs are often near many countries' popular tourist locations. Often they form part of the main attraction in these areas; this is especially the case with people wanting to scuba dive and snorkel. The reefs can also be used as fishing grounds for locals and have been the source of ornaments for tourists.

The coral reefs of Jamaica

Like many other popular holiday destinations, the Caribbean island of Jamaica has a large number of tourists each year who wish to explore the coral reefs off the coast. With the increasing demand for more hotels there is a danger of over using the coral ecosystems as a tourist attraction. This could lead to permanent damage of many unique ecosystems.

Assessment tasks

Sunshine Beach Project, Montego Bay, Jamaica, W.I.

Dear Sir/Madam

It gives me great pleasure to inform you that the development of the new 'Sunshine Beach Hotel' near Montego Bay, Jamaica, is nearly ready to begin.

The final stage is the Environmental Risk Assessment. This, as I'm sure you will know, is an evaluation of the potential environmental problems that this hotel may cause. As Jamaica's Minister for Tourism, I am sure that you are keen to see the development go ahead, but only after the project has been properly assessed for its impact on the local ecosystem.

Just to remind you of a few statistics concerning the hotel project.

- The 'Sunshine Beach Hotel' will have 500 bed spaces; it is expected that these will be full for nearly all the year.

- A new marina is planned for access to the hotel in Montego Bay. For it to work, a channel must be cut through the reef. It is thought that the new marina will bring in many more visitors and will handle up to 100 boats.

- The hotel complex will offer guests the chance to do any type of water sports activity they wish, ranging from jet skiing to scuba diving on the coral reef.

- It is hoped that the local community of Montego Bay will benefit from the jobs on offer in the new hotel and from the trade that the tourists will bring.

Yours faithfully,

Sarah King

Sarah King
Manager – The Sunshine Beach Project

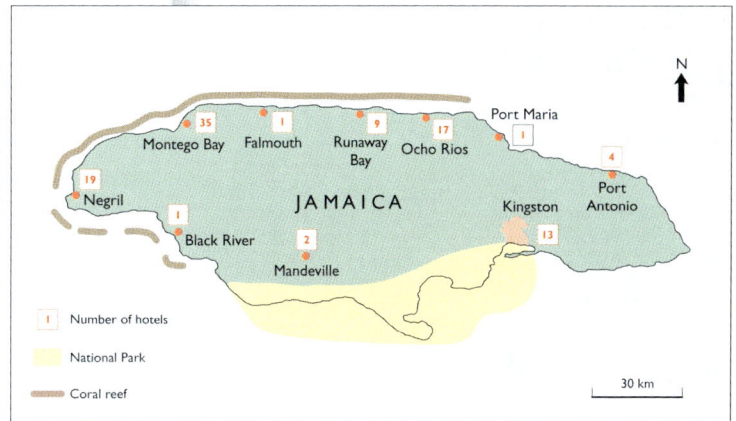

△ **Figure 5.30** Location of main tourist areas

▽ **Figure 5.32**   ▷ **Figure 5.31**

Impacts on:	Positive	Negative
People	1. Many locals will get jobs building the new hotel and working in it when it begins operating. 2. They will also have more tourists spending money in the local shops and businesses.	1. Much of the money made by the hotel will go back to the country where the multinational company is based. 2. Many jobs given to locals will be poorly paid.
The environment	**?**	1. An increase in tourists will mean more visitors to the reefs. This will mean more damage to the coral.

Assessment tasks

Your work as a researcher on the local newspaper has been to investigate the effects of tourism on Jamaica's coral reefs. You have collected the following pieces of information during your three week investigation. You know that all this information could be linked with threats to the coral reefs. It is not in any particular order and now you need to use it for your article. You will have to read each statement carefully to make sense of what you have found.

Frederick has high hopes for his illegal business operations.

Frederick remembers listening to his grandfather's stories of how their family has always fished the coral reefs for food.

Sergeant McDougal has had a long night. He has just locked up two fishermen who were caught using dynamite on the west coast of the island near Negril. This has been illegal for the past three years.

Sediment (silt and mud) blocks sunlight from the polyps, causing them to die.

James and Tina Douglas book a holiday on the Caribbean Island of Jamaica. They want to spend their time relaxing in the sun.

One of Ron's boats has had to return early. The divers could not see the coral as there have been too many boats at the reef today and the water is cloudy with sediment.

The coral reefs are under threat from locals who take pieces of coral to sell or make into jewellery. This is against the law, but difficult to prevent.

Ron Johnson has just bought three new boats for his 'Coral Reef Boat Tours' Business.

Coral reefs are very sensitive to changes in light, temperature, water clarity, salinity, and oxygen.

The power station near Montego Bay will have to be upgraded to handle the growing demand from the tourist facilities. This means even more warm water from its cooling system will flow out onto the reef.

John Whitehouse is a happy man. The Island's hotels have just increased their orders for his lettuces; he will have to contact his fertiliser supplier again.

The bank agrees to John's request for a business extension loan. He needs to buy more land beside the river.

Arthur, the building site manager for the Sunshine Beach Hotel, is having problems with the planned sewer system. The smell may cause complaints from the residents so Arthur thinks they need a longer pipe to take the flow further out into the sea by the reef.

Target task

1 Write an article for your newspaper explaining how you think the new hotel planned for Montego Bay might damage the coral. You may break this down into direct and indirect threats.

2 What could protesters do to explain why they don't want the development to go ahead?

3 Why might the local community want the hotel to be built?

Extension task

1 How could the developers reduce the risk of damage to the reef?
a In the short term
b In the long term

2 Should the people of Jamaica be put before the natural environment in the decision making process?

3 What other examples of new developments conflicting with existing ecosystems can you find, in the UK or in your local area.

◁ **Figure 5.33**

Review

Throughout this chapter you have looked at the theory of ecosystems and the issues affecting them.

This review page is designed to give you a chance to reflect on your understanding of this topic. The boxes at the top of each column represent your goal for each section of the topic. Start with the first column and read the statement in the bottom box. If you agree with it, move on to the box above. It is important that you are able to show your understanding at each level. An easy way to do this is to turn the statements into questions by switching the 'I can …' to 'can I…?'

If you want to move from one box to another but are not sure how to, ask your teacher for help.

› Issues to investigate further

• The Eden Project
• Over fishing in the North Sea
• **Eutrophication** – Wetlands' water quality and Agriculture
• The damage done by holiday developments in less economically developed countries (LEDCs).

To be able to explain how the complex number of links between parts of an ecosystem can be simplified, using systems thinking.	To be able to explain how climate affects the global distribution of ecosystems.	To be able to explain the importance of energy and nutrient flows within ecosystems.	To be able to explain how humans have exploited the resources of many of the worlds' ecosystems, often changing them forever.	To be able to explain how the future of many ecosystems depends on sustainable management of natural resources.
I can explain how the impact of a change to one part of an ecosystem can be seen tracked through the whole system.	I can explain the position of more than one global ecosystem, describing what the impact of global warming may have on their disributions.	I understand the idea of energy flows within food chains and I can explain how damaging the food web of an ecosystem can impact on its organisms at many different levels.	I can explain the short, medium and long term impacts of these acts of exploitation on the ecosystems.	I can give examples of ecosystems of different scales that have been affected by human activity and how a balance can be reached through sustainable management.
I can also name and describe how links in the ecosystem show processes at work between biotic and abiotic components.	I can also explain what biomes are and how their locations on the globe are related to the amount of energy they get from the sun.	I can also explain how organisms form more complex food webs within ecosystems.	I can also explain the reasons for the exploitation of natural resources of a named ecosystem.	I can also explain what sustainable management is and why it is important where ecosystems are under threat from humans.
I can give examples for an ecosystem of its inputs, flows, stores and outputs.	I can describe how temperature and precipitation affect the way plants and animals grow in ecosystems.	I can give an example of a food chain for an ecosystem, indicating which organisms live off each other.	I can explain what we mean by exploitation, giving an example of an ecosystem where this has happened.	I can explain what ecosystem management is and give examples of where it has been used to tackle problems caused by people.

6 The Environment

Is this a tale of two cities?

Figures 6.1 and 6.2 show the same area but 100 years apart. One of the few features that you can see in both is the church tower. What has happened to the landscape in the last 100 years?

▽ **Figure 6.1** Longton 1900

▽ **Figure 6.2** Longton today

1 Imagine you were living in Longton in 1900 and you were transported by a time machine to the same site 100 years later.

a What would you find strange about the new landscape?

b Would there be things about the view you would not be able to understand?

c What three questions would you like to ask a person who lives in the area today?

2 Now imagine you were transported back in time. You are able to walk through the streets (although people will be amazed at your clothes and trainers!) and study how people lived in those days.

a What surprises you about the changes that you see?

b What does not surprise you?

3 Look at the graph showing the different causes of death for the two periods (Figure 6.3).

a What causes of death were common in 1900, which are not today?

b How did the environment at home and at work affect the health of people in 1900?

c What causes of death are common now but not then? Has our way of life changed to make this so, e.g. smoking, unhealthy diet and lack of exercise?

4 People in 1900 had a shorter **life expectancy**. This is shown in the **population pyramids** (Figure 6.4) for 1900 and today.

a What three things do you notice that are different between the two?

b Try to give reasons for these differences. Is the environment to blame?

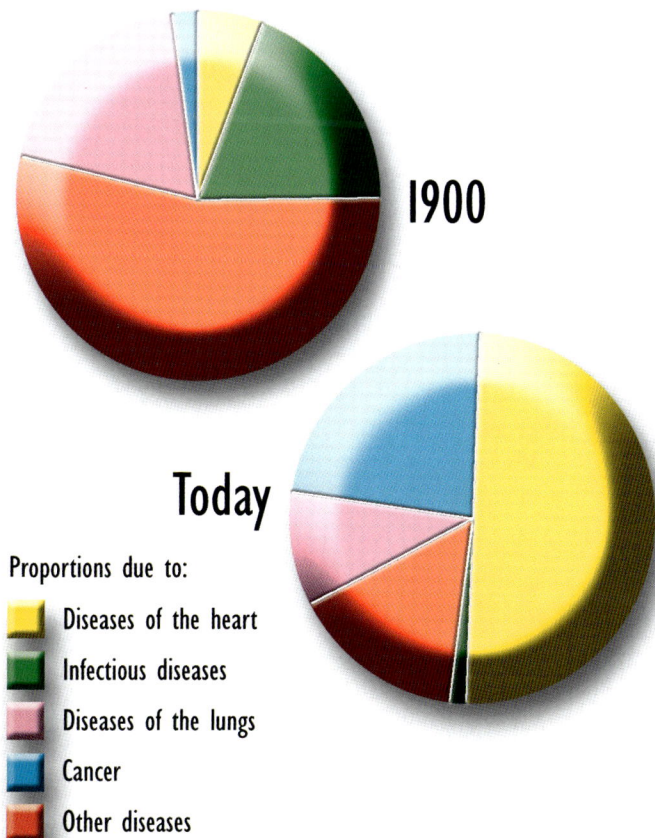

Proportions due to:

- Diseases of the heart
- Infectious diseases
- Diseases of the lungs
- Cancer
- Other diseases

△ **Figure 6.3** Causes of death in Longton, 1900 and today

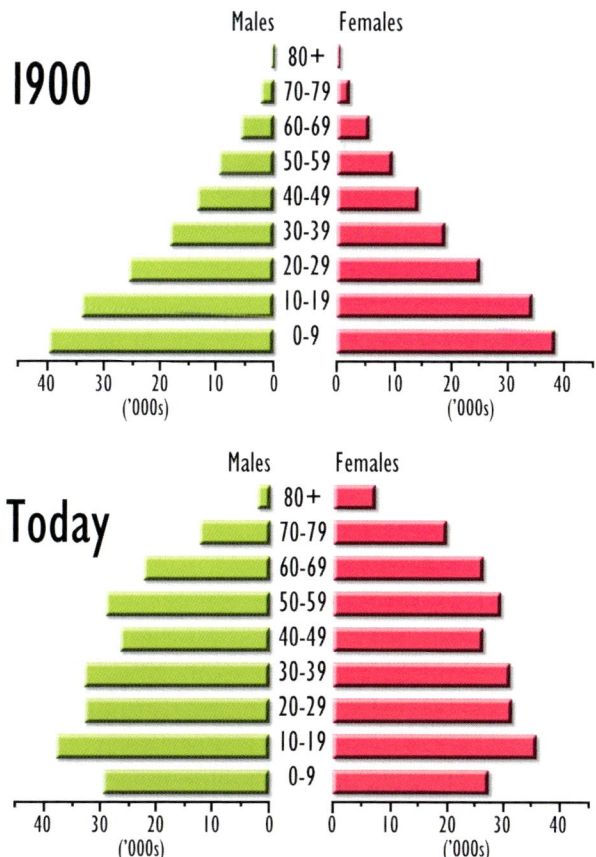

△ **Figure 6.4** Population pyramids for North Staffordshire, the area where Longton is situated, 1900 and today

What is the state of the planet?

The **environment** is another word for our surroundings. The natural environment is to do with air, land and water on our planet. Why are we so concerned about the global environment? Does it really affect us? Many of the environmental problems, such as air and water pollution, are to do with how resources on our planet are being used.

Activities

1 What do you think Figure 6.6 says about the use of world resources?

2 Is this situation sustainable?

3 What will happen as more countries get richer?

△ **Figure 6.5** Earth from space

△ **Figure 6.6**

The cost of using up resources is seen already with climate change, an increase in **pollution** and the **extinction** of plant and animal species. The following show how we are all in some way to blame for the state of the planet.

On a world scale

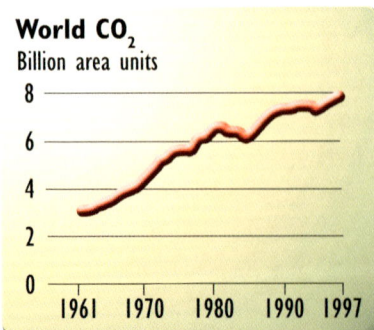

World CO$_2$ Billion area units

△ **Figure 6.7**

More carbon dioxide is given off as the world uses more cars and more industries use **fossil fuels**. Carbon dioxide (CO_2) makes up most of the greenhouse gas. More of this gas will create **global warming** with worldwide changes in sea levels through the ice cap melting and climate change. Forests can absorb carbon dioxide but we will need more trees to keep up with the increase of this gas. **Deforestation** is making matters worse.

Communities

Rich countries use up more resources and waste more. Something can be done about this, e.g. if the USA reduced its food waste by one-third, a further 26 million people could be fed!

Commerce and industry

Many products such as light bulbs and cars are often produced with a limited life cycle. The USA throws away 15 million working personal computers a year.

Consumption

We all need to consume to live but think about the energy used in transporting the products we buy, e.g. if the 800 000 tonnes of imported wheat used for European bread was produced in Europe, there would be an enormous saving in transport costs and less pollution.

Production

Making products will always use up the planet's resources whether **raw materials** or **energy**. **Globalisation** is also making big changes to the way products are moved around our planet. **Recycling** can reduce the impact on our planet, e.g. in Britain we could reduce carbon dioxide emission by 3 million tonnes a year if we recycled a third of plastic waste.

Waste

In a week we, in Britain, produce enough rubbish to fill Wembley Stadium. Richer countries produce the most waste per person per year, e.g. USA: 1 410 kg, UK: 525 kg.

Natural resources

Some resources are **non-renewable**, such as fossil fuels. Other **renewable** resources are also not being replaced quickly enough, e.g. the average size of sea fish is reducing as trawlers are going further in search of declining stocks.

Activities

Foundation

1 Where does the extra amount of carbon dioxide come from?

2 What will global warming do to sea levels and people living near the coast?

3 Give three examples, either from above or from your own knowledge, on how we could save the planet.

Target

4 How is global warming connected to fossil fuel use and deforestation?

5 Draw a table showing problems and solutions under the headings:
  • Communities
  • Commerce/Industry
  • Consumption
  • Waste
  • Production
  • Natural resources.

Extension

6 The 24th of November was declared 'Buy Nothing Day'. We all need to buy things to survive but can you suggest reasons why you would support this day? Would this apply to all of the people on our planet?

Is water quality in danger?

The Earth seen from space looks very much like a water planet. 70% of the globe is covered by water but only 3% of the world's water is fresh, most of which is stored as ice. Until recently many people thought that water supply was limitless. Today we recognise that about 2 billion people in the world actually suffer from water shortages. As population increases and countries industrialise, demand for water will increase and there will be an even greater threat of pollution.

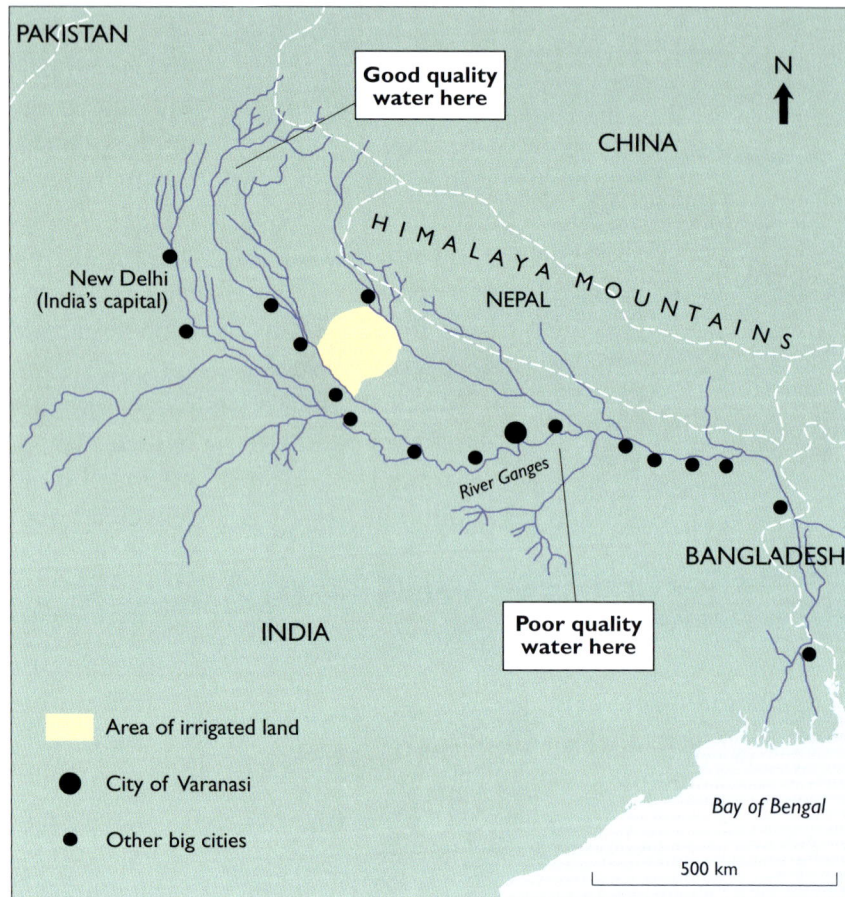

PAKISTAN

Good quality water here

CHINA

HIMALAYA MOUNTAINS

NEPAL

New Delhi (India's capital)

River Ganges

BANGLADESH

Poor quality water here

INDIA

Area of irrigated land

City of Varanasi

Other big cities

Bay of Bengal

500 km

△ **Figure 6.8** Map of the Ganges

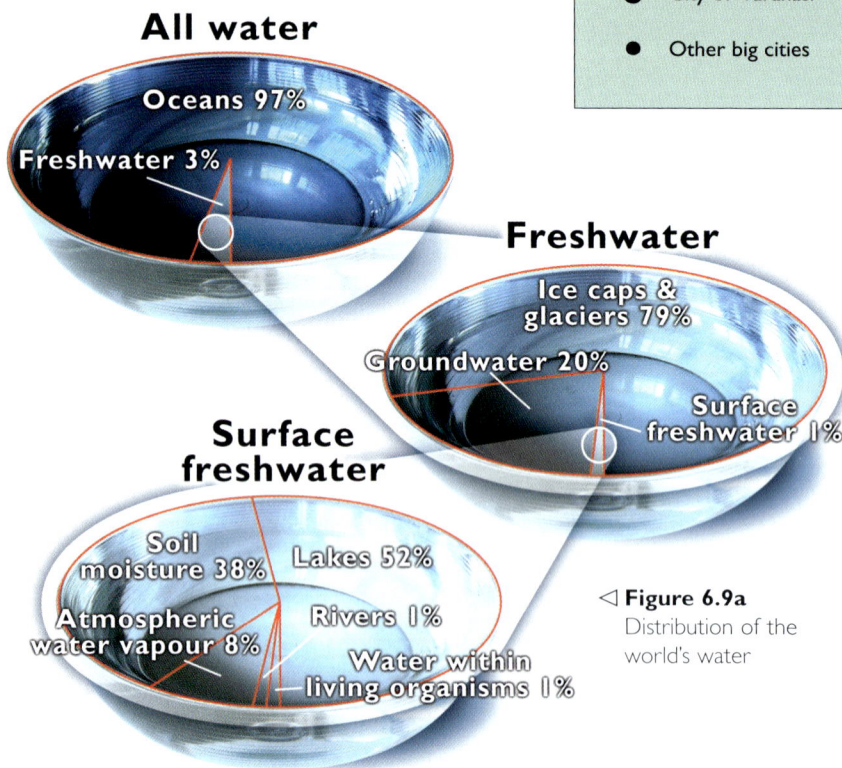

All water

Oceans 97%

Freshwater 3%

Freshwater

Ice caps & glaciers 79%

Groundwater 20%

Surface freshwater 1%

Surface freshwater

Soil moisture 38%

Lakes 52%

Atmospheric water vapour 8%

Rivers 1%

Water within living organisms 1%

◁ **Figure 6.9a** Distribution of the world's water

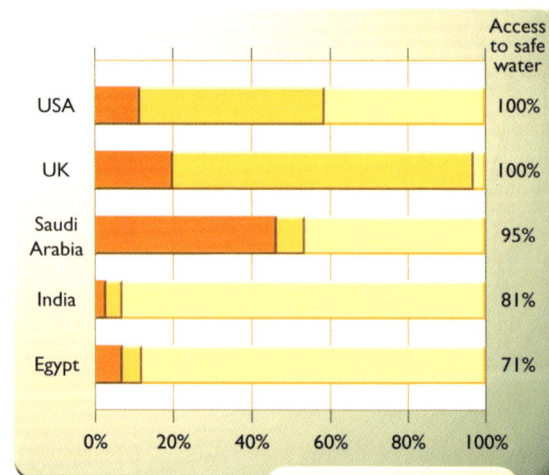

				Access to safe water
USA				100%
UK				100%
Saudi Arabia				95%
India				81%
Egypt				71%

0%   20%   40%   60%   80%   100%

△ **Figure 6.9b** How water is used in some countries and how safe it is

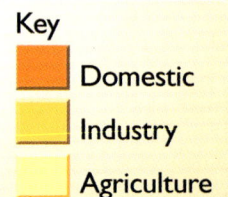

Key

Domestic

Industry

Agriculture

△ **Figure 6.10** Varanasi, a holy city on the banks of the Ganges

The Ganges River in India

Most of the river water comes from melting snow from the Himalayas and from rainfall during the monsoon; this is a season of heavy rain from June to September.

❯ The problem

The Ganges is sacred to Hindus, but for almost 600 of its 2 250 kilometres it is heavily polluted. How does it get so polluted? Pollution comes from three sources:

1 *Animal and human waste.* Many cities such as Varanasi have no sewage plants. 40% of India's billion population live in the Ganges Basin. Rainwater, especially during the monsoon season, flushes sewage directly into streams and rivers.

2 *Industrial waste.* Only 12 of the 132 industrial plants along the river have waste treatment plants. They make products ranging from paper, textiles, chemicals and leather.

3 *Agriculture.* Fertilisers and pesticides are washing off fields into the river. Water is also used in **irrigation**.

❯ What can be done?

India is a Less Economically Developed Country (LEDC) and so has fewer resources available to deal with this difficult problem. The country's population is also growing rapidly. Its birth rate of 26 per thousand per year is much higher than the death rate of 13. The country's economy is growing and many areas are developing with the location of new factories. However water quality and supply is a major problem for India. '*It's difficult to get water of any kind, let alone clean water. And the problem can only worsen,*' says an Indian pollution control expert.

The Ganges Action Plan was set up in 1985 with the aim of cleaning up the river. It cost over $300 million, much of the money coming from the **United Nations**. So far the project has revolved around building high technology sewage plants and has been a failure. Frequent power failure means that untreated sewage flows straight into the Ganges. In some places in Varanasi the concentration of pollutants is 340 000 times the permissible level! A local engineer says '*our technology is tried and tested and the work is partly successful. The next stage would make the river safe for bathing.*'

Amarnath, a pilgrim taking a dip in the river, ignored refuse, ash of cremated bodies and raw sewage. '*The river is pure. Germs cannot survive in its waters,*' he said.

> Is there a better solution to cleaning up the Ganges?

The head priest at Varanasi's Sankat Mochan temple is also a scientist and has suggested a plan to lay a 12 km pipeline to intercept sewage from Varanisi, diverting it to a series of ponds, which would clean the water using micro-organisms. The pipeline would not rely on electricity, but would use gravity to keep the water flowing. He said it would be cheaper because it is simpler.

△ **Figure 6.11** Pilgrims come to bathe in the Ganges. Pious Hindus believe that the waters will wash away their sins and being cremated there will guarantee an escape from suffering. An estimated 7.5 million people bathed in the Ganges on 14 January 2001, during Maha Kumbh Mela or Sacred Jug Festival

Activities

The Mayor of Varanasi is very concerned about pollution in the Ganges and wants the best for his people and pilgrim visitors. What should he do?

A Leave the situation as it is. India needs the money to develop industry. With more wealth India will eventually be able to afford a modern system in the future.

B Carry on with the sewage plant system. The technology is tried and tested. The general manager of the Ganges Pollution Control Unit, sums it up by saying 'Look how long it took to clean up the Thames in England and this is a much bigger river system.'

C Use micro-organisms in ponds to clear up sewage.

D Use a combination of **B** and **C**.

1 For each of the choices say what are the good and bad points.

2 Come to a decision about the best way to combat pollution of the Ganges.

a Why did you make this choice?

b What will be the impact of your choice on the environment and the life of people who live in the Ganges Basin?

ICT Activity

Produce a script for a two minute TV report on the problem of pollution in the Ganges. The Internet can provide opportunities for photographs to accompany your script. Here are some websites to start with:

www.csmonitor.com/durable/1997/10/29/intl/intl.6.html

http://ens.lycos.com/ens/jan2001/2001L-01-19-01.html

www.stfrancis.edu/ns/bromer/envhum/student8/

Are our oceans under threat?

The Oceans

Oceans have most of the world's water (Figure 6.9a) but due to their enormous size, we may think that the threat of pollution is less than it is for rivers. One way in which oceans are being polluted is from oil spillage. Figure 6.12 shows some of the major oil spills that have taken place in recent years. Many are not well-known because despite their large size, they caused little damage to the environment.

△ **Figure 6.13** The Exxon Valdez runs aground on Bligh Reef. Attempts to unload oil fail and large areas of the coast are polluted

▽ **Figure 6.12** The worst oil spills in the world. Note: Exxon Valdez is ranked at 34. It is listed because it is one of the most well-known

Rank	Name of ship	Date	Location	Tonnes spilt
1	Atlantic Empress	1979	off Tobago, West Indies	287 000
2	ABT Summer	1991	700 nautical miles off Angola	260 000
3	Castillo de Bellver	1983	off Saldanha Bay, South Africa	252 000
4	Amoco Cadiz	1978	off Brittany, France	223 000
5	Haven	1991	Genoa, Italy	144 000
34	Exxon Valdez	1989	Prince William Sound, Alaska, USA	37 000

Activities

1 Look at Figure 6.12. Find out where these oil spills are and locate them on an outline world map. Is this a global environmental problem? For all the major spills consult website: www.itopf.com/stats.html

2 Scientists have been studying the long term effects of oil spills. In Alaska a small rock affected by the Exxon Valdez spill has been photographed every year since the disaster that devastated so much of the wildlife there. To see the photos of this rock, consult website:
http://response.restoration.noaa.gov/photos/mearns/mearns.html

a Why do you think they are studying this rock?

b What will their results tell us about the long-term effects of the oil spill?

ICT links

Other sites of interest:
www.oilspill.state.ak.us
This has lots of information particularly about restoration work.

http://projects.edtech.sandi.net/encanto/disaster/
This site shows you how to put together a project on the Exxon Valdez oil disaster.

How much land is damaged?

When areas industrialise, land often gets damaged. This is a global problem and as the UK was the first country in the world to industrialise we have a long history of this problem. Waste material from industry such as coal mining can cause land to become derelict. Today much of this derelict land has been **reclaimed**. Other countries have also experienced these problems as they develop their industry.

Reclaiming damaged land in south Wales

In south Wales, some large industries have declined in recent years. Steelworks at Cardiff, Ebbw Vale and Llanwern (near Newport) have closed. The coal industry has nearly disappeared as well. This has left behind a problem of what to do with land that has been damaged.

Derelict buildings also result when industry closes. In Blaenavon, once a thriving coal and steel town, many shops have closed. Some new industry has been attracted here with government grants and a mining museum named 'Big Pit' has brought some jobs. However unemployment is still high.

△ **Figure 6.14** Llanwern steelworks before closure

◁ ▽ **Figure 6.15 and 6.16** New Tredegar. The mine was the heart of this mining village (left). In 1958, 1 375 miners worked underground as well as 321 on the surface. Today the mine is closed and the land has been reclaimed (below)

△ **Figure 6.18** The site of Ebbw Vale steelworks became a Garden Festival, part of which houses a shopping centre

▽ **Figure 6.17** Former miners guide visitors around 'Big Pit'

△ **Figure 6.19** Blaenavon, South Wales.

Activities

Foundation

1 What two industries have damaged land in South Wales?

2 What has happened to the land that was once damaged?

3 What new jobs have been created since the industry closed?

Target

4 What harmful effects have these two industries had on the environment?

5 How do you think local people were affected by the old industries closing down?

6 Why have some shops closed down in these areas?

Extension

7 Who do you think should pay for the damage done to the environment?

8 Explain why you would not be able to put a price on some of the environmental damage.

9 Jobs or environment? Can you reduce the conflict between creating jobs and damaging the environment?

A success story from Schlema, Germany

Schlema is a town in Saxony, an area of Germany that was once part of the former country of East Germany. It has suffered some very extreme problems of industrial pollution and dereliction but today Schlema has overcome these difficulties and is looking forward to a bright future. Try and piece together the story on the following pages.

Activities

1 Why is Hans Schmidt going back to Schlema? To find out, study the 12 statements on the following page. You may want to copy each of these onto a small piece of paper and put the story in order, or copy them out in the order you think is correct.

2 What jobs are now available for the young people of Schlema?

3 Would you like to live in Schlema? Explain your point of view.

◁ △ **Figure 6.20** Maps showing Schlema's location

In 1990 the mine was closed down following the Unification of Germany. The whole area was contaminated and had an uncertain future.

Hans went back to Schlema in the 1950s but was horrified by what he saw. Although he still had relatives there he vowed he would never go back.

In 1947 the occupying Soviet forces began mining for uranium for use in nuclear weapons.

The uranium mining destroyed the landscape.

In 2001 Hans was diagnosed as having rheumatism.

Schlema is a small town, South West of Chemnitz, in Saxony, a region in Eastern Germany.

In 1992 the decision was taken to decommission the mine works and rejuvenate the landscape.

Hams Schmidt's father moved to Schlema in the 1920s. He suffered from rheumatism and was attracted to the town by the health spa.

The redevelopment will include radon-containing medicinal springs used to treat rheumatism.

Schlema was a mining town where iron, copper, cobalt and silver were extracted.

Hans was born in Schlema and grew up in the town. He enjoyed living there until he left during the 1939–45 war.

In the 1920s a radioactive spring was discovered; this was used to treat rheumatism.

▷ **Figure 6.21** Hans Schmidt and family

△ **Figure 6.22** Reclamation in Schlema
Waste rock pile: 1960, 1993, 1993

△ **Figure 6.23** The uranium mine, now a tourist attraction

〉 **Have you solved the mystery?**

- Do you know why Hans Schmidt is going back to Schlema?
- Would you go back?
- What jobs are now available for the young people of Schlema?

ICT links

Websites to take you further:

http://forests.org/articles/reader.asp?linkid=8247
An article on how Germans are planting trees to reclaim the land

http://abcnews.go.com/sections/world/DailyNews/Radioactivespa010118.html
American report on the benefits of radiation at Schlema

www.bmwi.de/textonly/Homepage/download/english/wismut_e.pdf
This is an English language account of land reclamation in Saxony.

What a Choke!

Air pollution is a term used to describe any harmful gases in the air. Humans make this pollution by burning petrol or diesel in cars. Also harmful gases are given off by power stations and industry when they burn fossil fuels.

The car's the star – or is it?

Cars have become an essential part of our everyday life. However there are serious consequences for air pollution. Countries with high car ownership are already experiencing bad effects. Los Angeles, a large US city on the west coast, has very high car ownership. Despite attempts to reduce pollution by fitting cars with **catalytic converters**, air pollution is still a problem here. The city lies in a basin and so air pollution gets trapped.

△ **Figure 6.24** A photo from Los Angeles showing air pollution!

As countries develop and more vehicles become available, similar problems occur. Figure 6.25 shows the traffic-choked streets of Mexico City. Pollution from vehicles is by the mountains seen in the background. Rapid population growth and rising wealth has created these enormous traffic problems. This is despite the building of a subway system in 1970 and a ban on certain cars driving on a certain day of the week: days are chosen using the last number of the licence plate.

▽ **Figure 6.25** Traffic in the streets of Mexico City

	Number of cars per thousand population	Predicted population in millions 2025
USA	521	333
UK	359	60
Mexico	92	130
India	4	1330
China	3	1480

△ **Figure 6.26** Tables of number of cars/predicted population

Activities

Foundation

1 What physical factors can make air pollution worse in some cities?

2 What is being done to reduce vehicle pollution in Los Angeles and Mexico City? Do you think enough is being done?

Target

3 Study Figure 6.26. Why are there differences between the numbers of cars per thousand of the population?

4 Calculate the number of cars that may be found in each country in the year 2025.

Extension

5 Calculate the number of cars in India and China if car ownership increased fourfold by 2025. What effect would this have on the environment of these two countries?

Does air pollution mean wealth?

Air pollution was noticed in England in medieval London as more and more people used coal for their domestic fires. Figure 6.1 at the start of this chapter shows how polluted the air was during our **Industrial Revolution**.

The demand for electricity has meant bigger and bigger power stations. Figure 6.27 shows a large power station in the Czech Republic. The coal it burns is very harmful to the atmosphere. Increasingly new houses are being built near polluters (Figure 6.28).

Today many developing countries are polluting the air in the rush to get rich. India is now the sixth largest producer of **greenhouse gases**. Air pollution will create 2.5 million early deaths.

In China development is also having its effects on air quality (see Figure 6.29). The north-east city of Benxi is one of China's centres of steel and cement production. Here 7 million tonnes of poor quality coal, high in sulphur, is used for electricity generation, industry and domestic use. There are few controls from the government. Attempts are now being made to reduce pollution from the steelworks and use waste gas to heat

◁ **Figure 6.27** Pollution from electricity generation, Czech Republic

◁ **Figure 6.28** Demand for new housing in an area, despite being close to a polluter

◁ **Figure 6.29** Heavy industrial pollution in the Liaoning Province of north-east China

Common air pollutants

Smoke	This is easily seen as it is made up of small solids.
Sulphur dioxide	This is a colourless gas mainly coming out of **thermal power stations.** When mixed with water, **acid rain** is produced.
Carbon monoxide	This gas has no colour or smell. It is produced by road transport.
Hydrocarbons	Produced when petrol is not fully burnt. Helps the formation of **smog**.
Nitrogen oxide	Given out by vehicles and power stations. High amounts in cities.
Ozone	Main gas in **photochemical smog** produced when nitrogen oxides and hydrocarbons react in sunlight.
Particulates	Small bits of solid or liquid, e.g. soot, fumes carried in the air. Produced by vehicles, industry and domestic fires.

△ **Figure 6.30** Table of common air pollutants

Activities

1 Why does development bring with it air pollution?

2 Why are developing countries now suffering from polluted air?

3 When countries get more developed the air pollution often gets less. Can you explain why?

4 Study Figure 6.30. Design a leaflet that would help young people understand the different types of air pollutants. You should include images as well as key facts.

Should we think globally, act locally?

Many people often wonder whether they can do anything about global environmental problems. If it is global, how can we help? There are some things that can be done as individuals and by local authorities.

In 1992 there was an important United Nations conference in Rio de Janeiro. Over 150 countries attended, including Britain. This conference has been called the Earth Summit. All the countries signed up for **Agenda 21**, which sets out how countries can work together to create development, which is more sustainable. Although Agenda 21 is mainly concerned with climate change and deforestation, it suggests that local action can also do its bit. It calls for councils to work with local people to develop plans for sustainability where they live.

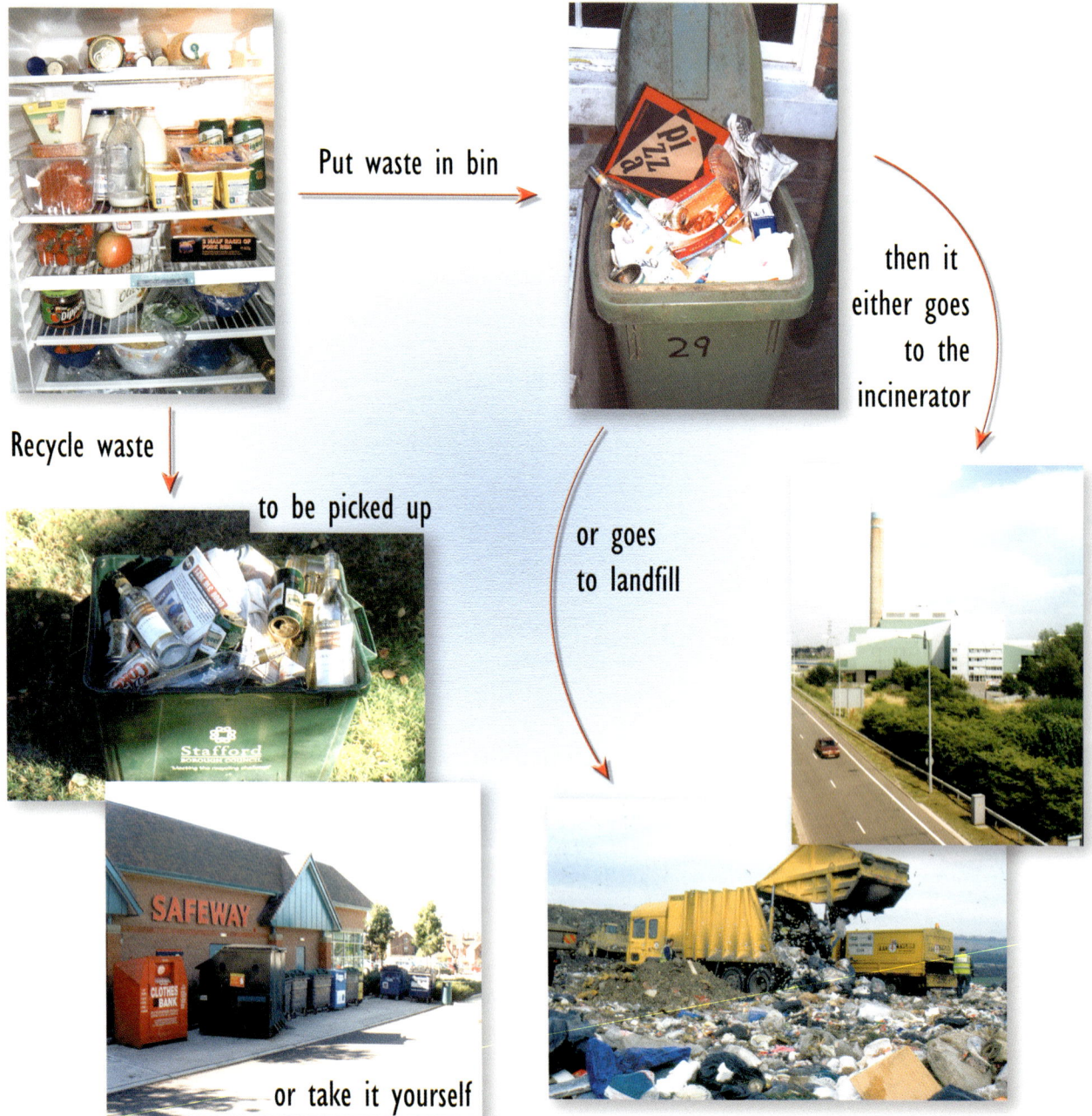

Put waste in bin

then it either goes to the incinerator

Recycle waste

to be picked up

or goes to landfill

or take it yourself

△ **Figure 6.31** Here are some of the possible routes of household waste

▷ **Figure 6.32** Recycling Green waste

FACT FILE:

Recycling: The idea is to recapture materials that would otherwise go to waste.
Green boxes are being used by some councils who pick them up from your home. People also take their materials to an area where containers are available for recycled products. Ask yourself whether a trip to recycle saves resources if you are driving your car there.

Landfill: This is putting waste into holes in the ground. There are problems of waste gas, **leaching** of chemicals into water supply and using up land. Some scientists have said that mothers living close to a landfill site have an increased risk of having babies with birth defects.

Incineration: Burning the rubbish. One of the photos in Figure 6.32 shows an incinerator in Stoke-on-Trent. Here rubbish is brought from the city and burnt. 12.5 MW of electricity is also generated. Local people worry about possible harmful effects from this burning.

Activities

Figure 6.31 shows you where your waste might end up. A meeting has been called by your local council to see how far they have got with Agenda 21. You are a reporter for a local newspaper and are keen to write up a story on this issue. Your feature should fill one page of a newspaper.

It should have:
  a an eye-catching headline
  b an introduction explaining what Agenda 21 is
  c a report on what is being done
  d comments on its success here
  e the pitfalls
  f a conclusion – perhaps the old Chinese proverb, 'the journey of a thousand miles begins with one step'.

(Some useful websites are listed on page 122.)

Milk bottle recycling competition
Win a portable TV and playstation

There are 5 prize packs to give away including a £50 shopping voucher and a T-shirt made from recycled plastic. One lucky person will also win a portable colour television and a playstation!!

To enter, simply write your name and telephone number / contact details on each milk bottle, remove the caps, squash the bottles and put them in this bank.

This competition is run by Recoup. Full competition details and lots more information about plastic bottle recycling is available on our website at www.recoup.org

FRESH MILK

△ **Figure 6.33** Can competitions encourage more recycling?

Target tasks

Read the information about the ship *Khian Sea* and how it spent years looking for somewhere to dump its waste (known as garbage in the USA).

1 What waste were they trying to get rid of?

2 Why did they not try to dispose of it in the USA?

3 Why would countries accept waste in the first place?

4 Why would a country not want somebody else's waste?

5 Why do you think they stopped the ship unloading in Haiti?

6 What happened to the waste in the end?

7 Can we do anything to stop something like this happening again?

8 Although laws have been made to try to stop **developed countries** dumping **toxic waste** in **developing countries**, it still goes on. Why?

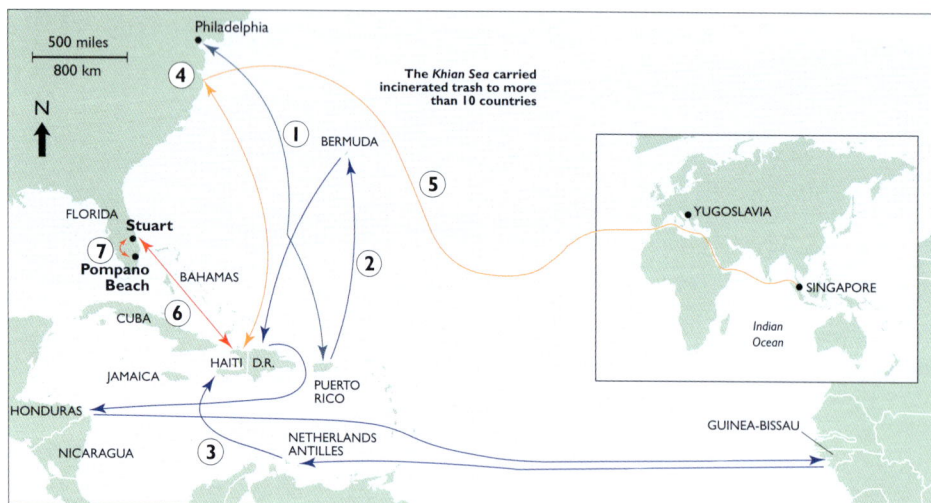

The *Khian Sea* carried incinerated trash to more than 10 countries

The trash's journey to South Florida

① Sept. 5, 1986
The *Khian Sea* leaves Philadelphia for the Bahamas with 14,355 tons of incinerator ash. Officials in the Bahamas reject the shipment before it arrives, forcing the ship to change course.

② Sept. 1986–Aug. 1987
The barge is turned away from Pueto Rico, Bermuda, the Dominican Republic, Honduras, Guinea-Bissau and the Netherlands Antilles.

③ Dec. 1987–Feb. 1988
The ship arrives near Gonairves, Haiti. The crew unloads about 4,000 tons of ash officially describes as 'soil fertilizer' before being stopped. The government orders them to reload and depart, but they leave the ash on the beach.

④ March 1, 1988
The *Khian Sea* arrives in Delaware Bay to return the remaining ash, but it is rejected. The ship leaves for Africa.

⑤ Aug.–Nov. 1988
The ship, renamed *Felicio*, appears in Yugoslavia for repairs and is escorted away. It reappears near Singapore with a third name, *Pelcano*, but no ash. The captain testifies that it was dumped in the Atlantic and Indian oceans.

⑥ Spring 2000
The ash left in Haiti is picked up and brought to Stuart. It sits in the St. Lucie River until a home if found.

⑦ Thurday, 26 January 2001
The state agrees to bury the ash in the landfill near Pompano Beach.

△ **Figure 6.34** Unwanted waste heads for Brownard on the ship Khian Sea. NGOs such as the Basel Action Network are making big efforts to stop rich countries dumping their waste in poor countries. Look up their website for more information: www.ban.org

At the Earth Summit in Rio people around the world were asked to consider the sustainability of development. We need to consider whether developments in our local area do really help the global environment.

One local issue that you will probably have first-hand knowledge about concerns itself with **greenfield** or **brownfield** sites. Many towns are expanding and there is an enormous demand for housing land. Study Figures 6.35–6.39 and think about the work you have done on this topic. Then, answer questions 1–3 below.

1 Draw a poster to show the advantages and is disadvantages of developing brownfield and greenfield sites.

2 Consider Agenda 21. How sustainable are both types of development? Hints! Remember that in clearing brownfield sites you will be using more energy. But once developed people will be closer to town and will use less fossil fuel in transport.

3 Can you suggest better ways of incorporating sustainable development into any building, e.g. energy saving devices?

△ **Figure 6.35** Cows graze on a greenfield site, but not for long

△ **Figure 6.36** New houses built on farmland (a greenfield site). The rural-urban fringe is under threat

◁ **Figure 6.37** An old shoe factory built in Stone, Staffordshire. Today its location, size and shape is less attractive to industrialists

▽ **Figure 6.38** The factory is demolished and a number of 'town houses' are built on this brownfield site

▽ **Figure 6.39** The site is cleared

Assessment tasks

ICT activities

Environmental report for London

Figure 6.40 shows how many tonnes of resources London uses and wastes.

1 You have been asked to prepare some graphs to help the Mayor of London present a report to an environmental conference. Think of a catchy title for the display and include some written comments to go with it.

2 Design a poster using your computer to highlight the problems of London's high use of resources and waste. This statement may help your understanding of this issue: *'Cities occupy 2% of the land surface of the world but use 75% of the world's resources.'* Your poster should try and suggest a solution.

▽ **Figure 6.40** What London uses and wastes

Uses	Tonnes per year
Fuel (oil equivalent)	20 000 000
Oxygen	40 000 000
Water	1 002 000 000
Food	2 400 000
Timber	1 200 000
Paper	2 200 000
Plastics	2 100 000
Glass	360 000
Cement	1 940 000
Bricks, blocks, sand, tarmax	6 000 000
Metals	1 200 000

Waste	Tonnes per year
Carbon dioxide	60 000 000
Sulphur dioxide	400 000
Nitrogen oxides	280 000
Sewage	7 500 000
Industrial waste	11 400 000
Household	3 900 000

△ **Figure 6.41** Large cities such as London create large quantities of industrial waste

Checking out websites

There are many websites on environmental issues. Here are some that can show how *you* can help the environment. For each website, say what you liked about it, what you disliked about it and give it a score out of ten.

Website Warning!
Try and think why websites are there. Who pays for them? Do they give all points of view? Are they truthful?

www.wastewatch.org.uk/
Go to the Kids Home Page

www.foe.org.uk/campaigns/industry_and_pollution/factorywatch/

Find out about pollution near you. Just type in your postcode, and a map with a list of polluters will be shown.

www.doingyourbit.org.uk
How you can help the environment

www.globalactionplan.org.uk
Small changes can make a difference

www.aeat.co.uk/netcen/airqual
Information on air quality in Britain. Find the education section

ICT Activity

Some websites can give you details of your local area once you give them your postcode details. Factory Watch is found on the Friends of the Earth website (see table on previous page for web address).

Figure 6.42 is from this site and shows the Liverpool area of North West England. It is an industrial area with many chemical plants along the banks of the Mersey Estuary. One plant is an oil refinery belonging to Shell. Its postcode is L65 4HB. By typing this postcode in to 'Factory Watch', details of chemicals put into the air will be listed.

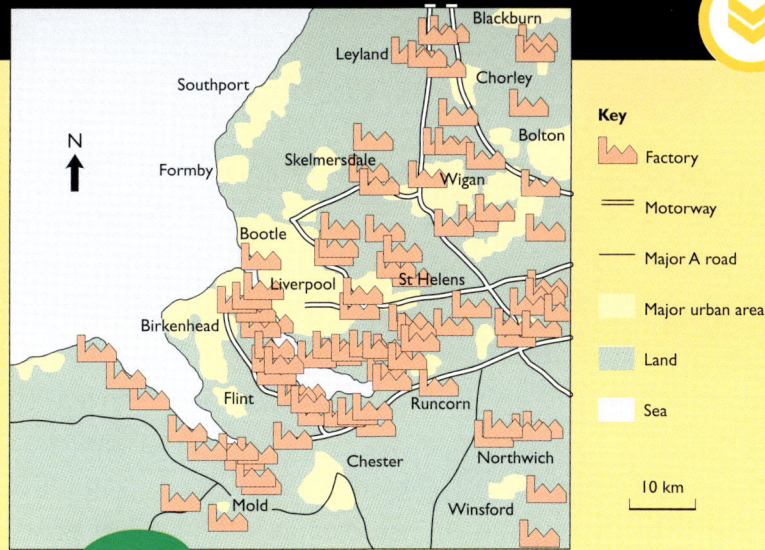

△ **Figure 6.42** Map of Liverpool area of North West England, showing chemical plants

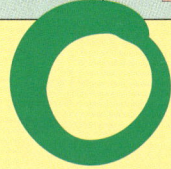

Friends of the Earth   ◁ **Figure 6.43**

Review

Environment Quiz

Choose the correct answer:

1 Sustainable development is:
  a consuming today with tomorrow in mind
  b taking as much from the earth as possible, regardless of the consequences

2 Agenda 21:
  a is a proposal to raise the drinking age to 21
  b was agreed at the Earth Summit in 1992. It encourages local action on sustainability

3 Global warming is mainly caused by:
  a producing greenhouse gases
  b nuclear power stations

4 Which are not examples of renewable energy?
  a coal, oil, natural gas
  b wind, solar, tidal power

5 LEDC countries are becoming more polluted because:
  a population is falling and industry is closing down
  b population is rising and more industry is forming

6 Schlema is:
  a a town in Saxony where land has been reclaimed
  b a device to control factory emissions

7 Smog is:
  a mainly caused by emissions from cars
  b mainly caused by barbecues

8 Landfill is used to:
  a dispose of waste
  b generate electricity

9 Recycling aims to:
  a recapture materials to use them again
  b build cycle lanes on busy roads

10 A brownfield site has:
  a been ploughed and so looks this colour
  b has been built on before

If you're not really sure – the answers are on page 129!

Glossary

Abiotic

A non-living thing such as soil

Acid rain

Rain having sulphuric and nitric acid strong enough to harm vegetation and freshwater life

Agenda 21

An agreement about environmental sustainability reached at the Rio Summit in 1992

Amenities

Something that benefits the residents of an area. This could include a swimming pool or a good shopping area

Ancestors

People from whom your family is descended

Annotate

Add notes to explain

Biomes

Global scale ecosystems

Biotic

A living thing, such as a rabbit

Birth rate

The number of births per thousand people in a year

Brownfield

An area for development that has already been built on

Car pool

The sharing of cars in journeys to work

Carnivore

An animal which eats meat.

Cartographers

People who design and draw maps

Catalytic converter

A device fitted to a car exhaust to reduce the emission of poisonous gases

Census

An official count of population data. The UK has a census every ten years, e.g. 1991, 2001, 2011

Central Business District (CBD)

The centre of towns and cities where shops and offices are found

Civil War

A war between people from the same country

Components

Interconnected parts of an ecosystem

Conflict

A struggle between people when they have different ideas about how to use the same resources

Connectivity

The ability to join

Conservationist

Someone who wishes to protect nature and the environment

Constructive plate boundary

Where new crust is formed at a mid ocean ridge

Consumer

Something which eats something else. In a food chain, the creature which eats the plants is referred to as a primary consumer; the creature which eats the primary consumer is called the secondary consumer and so on.

Conurbation

A large built up area made up of groups of towns

Counter-urbanisation

The movement of people from housing in towns and cities to the countryside

Country

The area of land that a state occupies

Cremation

The burning to ashes, usually of human remains

Cultural

Concerning the collective beliefs and traditions that influence the behaviour of a particular group of people

Death rate

The number of deaths per thousand people in a year

Decomposer

An organism which lives off the remains of dead plants and animals, speeding up the process of decay, e.g. bacteria or fungi

Deforestation

Cutting trees down without replanting

Democracy

Government by elected representatives of the people

Derelict

Unused and abandoned land

Destructive plate boundary

Where crust is destroyed as it is subducted back down into the mantle

Digital divide

The gap between people and countries who have access to technology and those who don't

Directory

A book published in the 19th- and early 20th-centuries listing the residents of an area and their occupations

Distribution

The spread of something throughout an area

Economic

The distribution of wealth in a country; also a description as to what businesses created that wealth and how it is consumed

Energy

Force needed to do work – can be renewable or non-renewable

Entrepreneurship

The ability to develop a new idea, particularly as a business, often at personal financial risk

Environment

Our surroundings – natural environment refers to land, water and air

Eutrophication

What happens when nutrients from fertilisers are washed into areas of freshwater, causing plants like algae to grow rapidly. As they grow all the oxygen in the water is used up, leaving none for other organisms in the ecosystem. The result can be dead and lifeless lakes and rivers

Evolve

To grow and adapt to an ecosystem

Exponentially

An increasingly steep rate of increase

Extinction

Complete destruction of animal and plant species

Faults

Weaknesses in rocks which can be easily affected by earthquakes

Federal

United as a nation, but with independence on local matters

Food chain

A sequence of living things which feed off each other, e.g. a wolf eats a reindeer which eats grass

Food web

A set of interconnected food chains, such as a food web of all living things in a hedgerow

Glossary

Fossil fuels
Created in past geological periods from hydrocarbons, e.g. coal, natural gas and oil

Friends of the Earth
A non-governmental organisation of people who support the environment and conservation

Garden Festival
An attempt to reclaim derelict land in Britain from old industrial sites

Global
Worldwide or whole world distributions, e.g. earthquakes, population densities

Global warming
The heating up of the earth's atmosphere

Globalisation
Companies operating world wide, e.g. Coca Cola

Greenfield
A site that has not been built on before

Greenhouse gases
A gas which contributes to global warming such as carbon dioxide, methane, ozone

Hamlet
A small rural settlement without services

Herbivore
An animal which eats plants

Hotspots
Weaknesses in the earth's crust which allow magma to erupt from the mantle

Immigrant
A person who moves into a country permanently

Incineration
Burning to ashes.

Individualism
Free action by any one person

Industrial Revolution
The change a country goes through to become an industrial nation

Informal job
One which is 'unofficial' such as shoe-shining, selling food from street stalls

Innovation
Change

International migrants
People who move from one country to another. Some move for a better life, others are forced

Irrigation
Supplying land with water artificially, e.g. through channels

Lahar
A mudslide on the slopes of a volcano caused by ice melt or rain saturating volcanic ash

Landfill
A site used to bury waste, usually a hole in the ground

Lava
The name given to molten rock when it erupts from a volcano

Leaching
The washing out of material by water passing through

LEDC
Less Economically Developed Country

Life expectancy
The average number of years someone is expected to live

Local scale
A unique area about the size of a school catchment area

Magma
The name for the molten rock of the mantle

Mantle

The part of the earth, below the Crust, that is liquid, molten rock

Map symbols

A sign on a map for a feature, e.g. a station is a red circle on some maps

Marginalised

On the edge of something or somewhere

MEDC

A More Economically Developed Country

Mercalli Scale

A scale describing the effects of earthquakes

Migrate

A general term to describe the movement of population. This can be either permanent or temporary

Nation

People having a common language, history etc; overseen by one government

National scale

A whole political unit, e.g. the United Kingdom or a country, e.g. France

Natural hazard

Physical processes or events that have the potential to cause harm to life or property

NGO

Non-governmental organisation

Non-renewable

Something that cannot be made again and is finite, e.g. fossil fuels

Open source software

When the thinking behind a program is freely available to all

Partial

Not total or complete

Pastoral

Farming with animals

Pattern

The way in which features occur or are arranged

Peninsular

Piece of land almost surrounded by water

Personal scale

An individual's immediate surroundings, i.e. the space close to their body

Photochemical smog

Created when exhaust fumes and factory emmissions combine with sunlight to produce ozone

Plateau

A large fairly flat area of upland

Pollution

When something is contaminated by human activity

Population density

A measure of how crowded an area of land is, normally the number of people per square kilometre

Population pyramid

A graph showing the age and sex of a country's population

Potential

Capable of developing

Producer

Plants which make (produce) energy by photosynthesis

Province

An adminsitrative division of a country

Pull factors

Reasons why people are attracted to other places and move there, e.g. to get a better job

Push factors

Reasons why people leave places, often from villages, to live in towns and cities, e.g. through lack of land

Glossary

Raw materials

Anything that can be processed to make a product, e.g. iron ore is one of the raw materials to make steel

Reclaimed land

Either damaged land restored to a useful state again or land converted to agriculture from the sea, marsh or desert

Recycling

Re-using materials which would normally be thrown away, e.g. paper, drink cans

Region

An area that may be made up of part of one or several countries, that has either physical or human feature(s) unique to it

Renewable

Resources which can be used repeatedly such as hydro-electric power (HEP),wind, wood

Residential

Areas of housing

Ricter Scale

A scale for measuring the force of earthquakes

Rural

The countryside

Rural urban migrant

A person who moves from the country to a town or city

Rural–urban fringe

The zone on the edge of a urban area (town or city) which comes into contact with a rural (country) area

Seismic activity

Movement of the earth's crust detected by seismometers

Seismometer

A sensitive instrument for measuring the shockwave created by tectonic activity

Settlement

Any form of habitation from a single house to a large city

Shanty town

In a city of a developing country where houses are built illegally in a makeshift way

Smog

A mixture of smoke and fog

Spa

A place where mineral springs are found, named after Spa in Belgium

State

An area of land whose people have an independent government

Subduction

The process that causes dense tectonic plates to dip below less dense plates when they are forced together

Subsidy

A grant of money that is given, e.g. to provide more buses and routes in a rural area

Subsistence

A type of farming where all the produce goes to feed the family and none is for sale

Sustainable

Using resources in a way that will allow more use in the future, e.g. recycling cans or saving energy

Sustainable development

To use the resources of an ecosytem in a way that ensures they are replaced for future generations.

Technological

Scientific, industrial and mechanical

Tectonic activity

Events caused by the movement of the earth's surface, e.g, earthquakes and volcanoes

Tectonic plates

Parts of the earth's crust that float on top of the liquid mantle

Telework

To work at home using telephones and computers to contact others

Thermal power station

Makes electricity by heating water into steam which turns generators

Toxic waste

Waste that is harmful or deadly

Tsunami

A large sea wave generated in the oceans by earthquake or landslides

United Nations

An organisation of most countries in the world. It tries to promote peace, security and international co-operation

Urban

Of towns and cities

Urbanisation

The process by which an area becomes built up; an extension of an area of houses, shops, offices and industry

Vegetation

Plants

Village

A small rural settlement larger than a hamlet often with a church, shop and public house

Vulcanology

The study of volcanoes

(Answers to Environment Quiz on page 123: 1a; 2b; 3a; 4a; 5b; 6a; 7a; 8a; 9a; 10b)

Index